Port-Focal Logistics and Global Supply Chains

Port-Focal Logistics and Global Supply Chains

Adolf K. Y. Ng
University of Manitoba, Canada

and

John J. Liu
City University of Hong Kong, China

palgrave
macmillan

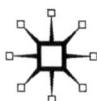

First published 2014 by
PALGRAVE MACMILLAN

Palgrave Macmillan in the UK is an imprint of Macmillan Publishers Limited, registered in England, company number 785998, of Houndmills, Basingstoke, Hampshire RG21 6XS.

Palgrave Macmillan in the US is a division of St Martin's Press LLC, 175 Fifth Avenue, New York, NY 10010.

Palgrave Macmillan is the global academic imprint of the above companies and has companies and representatives throughout the world.

Palgrave® and Macmillan® are registered trademarks in the United States, the United Kingdom, Europe and other countries

ISBN: 978–1–137–27368–0

This book is printed on paper suitable for recycling and made from fully managed and sustained forest sources. Logging, pulping and manufacturing processes are expected to conform to the environmental regulations of the country of origin.

A catalogue record for this book is available from the British Library.

A catalog record for this book is available from the Library of Congress.

Contents

List of Figures

List of Illustrations

List of Tables

Preface and Acknowledgments

Ranging from the rapid increase in international trade and the adoption of containers in most cargo transportation in the 1950s, to the end of the Cold War in the early 1990s and, more recently, the downturn due to the global financial crisis in 2008, there are few doubts that the world's economy has experienced fundamental changes in the past half century. Many of these changes not only were unprecedented and far-reaching, in both positive and negative ways, but also affected all walks of life. Being the arteries of the global economy and international trade, unsurprisingly, the shipping, port and logistics sectors continued to evolve and adapt to new challenges and phenomena. Initially, shipping, ports and logistical services were separate, and segregated, economic activities. Nowadays, along with the inevitable trend of globalization, they have evolved and increasingly combined into an integrated service profession – a process that has accelerated in the past decade. Being the nodal points, ports, which include seaports, airports, dry ports/inland terminals and other logistical centers, will play increasingly important roles in determining the success of global supply chains in both developed and developing economies.

In the past decades, substantial research has been conducted on the transformation of ports, logistics and supply chains. However, many of these have adopted a similar analytical path, evolving from the optimization-based disciplines, notably operations management and management science, especially in relation to supply chain management. On the other hand, there are a considerable number of studies describing how ports and logistics have evolved: for example, the former's infiltration in hinterlands, port centric logistics and collaboration between proximate (in many cases, competing) ports. However, most are rather piecemeal and based on geographical areas strongly influenced by local characteristics, situation and culture. Indeed, while we were reviewing previous studies, two questions often came to our minds: why were the existing theories and conceptual frameworks often restricted to certain cases, and why has it often been difficult to apply them elsewhere? It seems that the future of ports, and their potential roles in logistics and supply chains, still has myriad questions yet to be resolved satisfactorily, and the urgent need to fill this research gap is clear.

Such questions inspired us to initiate this book project in early 2012, and provided us with the motivation to finish it. This book grew from our research work and our highly diversified academic backgrounds. One of us trained as a human geographer in the UK and the other as an industrial engineer in the US. We have spent all of our careers so far in business schools and economics faculties in North America, East Asia and Western Europe, focusing on the study, research and teaching of the history, evolution and future development of (maritime) transportation, logistics and supply chain management. As far as we know, there are not many similar books in the international market written by authors with such diversified backgrounds and who have the competency and experience to look at the issue from so many academic and geographical angles. Throughout the past years, we have undertaken various advisory roles, and thus have provided key strategic advice to industries and (inter-)governmental organizations in the transport, logistics and supply chain sectors, including the United Nations (UN), the European Commission and the Government of the Hong Kong Special Administrative Region (HKSAR), to name but a few. These experiences ensure that we possess substantial first-hand information on the development and transformation that the sectors, and indeed the world, have experienced in the past decades. With such backgrounds, both of us have the vision and passion to share our experiences, research findings and ideas with academic colleagues, policymakers, industrial practitioners and students currently studying in universities and other tertiary institutions who will, no doubt, play pivotal roles in shaping the future of transportation, logistics and global supply chains.

This book is a valuable referene to all researchers and students focusing on ports, transport, logistics and supply chain research and studies. It is especially relevant for postgraduate research students who are focusing on the following subjects: maritime transport and logistics management; international trade and shipping; port economics, policy and management; tranport infrastructure planning and development; and transport and economic geographies. Also, it serves as an ideal companion to all policymakers and practitioners within the shipping, port, logistics and supply chain sectors who are eager to thoroughly understand the contemporary development, trends and challenges of ports, logistics and global supply chains, in both developed and developing countries and regions.

Moreover, we would like to emphasize that the content of this book comes from various countries and regions in both the developed and the developing world, including Brazil, China, Hong Kong, India, the Netherlands, South Korea and Singapore. In the past few years, we have undertaken various field trips to these countries and conducted semi-structured, in-depth interviews with many relevant personnel. They included reputable scholars and key decision-makers within the transport, logistics and supply chain sectors, notably senior policymakers and industrial practitioners. With such input, we are confident that our analysis, discussions, conclusions and ideas, as reflected in this book, are scientific, objective and highly applicable in a global context. We strongly believe that this book comes at an opportune moment when the world desperately needs to identify appropriate strategies and effective solutions for the current and future development of ports, logistics and supply chains. It comes at a time when these issues have become increasingly uncertain, and sometimes even controversial, and are not helped by the rise of various exogenous challenges, such as safety, security, climate change and sustainability. Indeed, the core concept introduced in this book – port-focal logistics – is not only effective in explaining the current direction of development of ports, logistics and supply chains, but also a highly appropriate way forward to develop efficient port-integrated logistical systems around the world.

It is to be noted that some of the contents of this book consist of our previous research work, covering a wide range of interrelated topics and featured (in different versions) in various forms of publication, e.g. scholarly journals, professional journals, proceedings of international conferences and a doctoral thesis. For further details, readers should refer to Chapter 1.

We would like to take this opportunity to say 'thank you' to our research assistants, who have helped taking some of the burden off our shoulders, namely Ellie Chow, Cherry Man and Flavio Padilha. Also, there are many colleagues who have provided useful information and suggestions that have substantially improved this book's content. They include Ismail Cetin, César Ducruet, Oleg Golubchikov, Girish Gujar, Peter Hall, Daisuke Ikemoto, Chung-Lun Li, Athanasios Pallis, Xinyu Sun, Jose Tongzon and Jia Yan. We would also like to thank all our interviewees, whose information and insights were critical in enhancing our understanding of the topic. Indeed, many of the source materials were unpublished (and confidential) and would be very difficult, if not impossible, to obtain otherwise. Of course, the faith of the publisher, Palgrave Macmillan, in our project is highly appreciated, while we gratefully acknowledge the generous financial support of the University of Manitoba's VPRI and the I.H. Asper School of Business Research Funds (314942). On a personal note, the first author would like to express his warmest gratitude to his parents and fiancée, who have provided him with unconditional support throughout the years – through countless nights of writing and continuous periods away from them overseas – and helped him overcome so many obstacles when developing his academic career. This book is dedicated to them.

The idea that this book examines will remain pivotal in shaping the appropriate development of ports, logistics and supply chains that most of us hope will ensue. It can, and will, offer paradigm-shifting insight into one of the most serious problems the world has ever faced, and is likely to face now and in the future. We are confident that its informative content, quality and objectivity will secure critical acclaim from scholars, policymakers and industrial practitioners alike.

Adolf K.Y. Ng
Winnipeg, Manitoba, Canada
John J. Liu
Hong Kong, China
September 2013

About the Authors

Adolf K. Y. Ng is Associate Professor in Transport, Logistics and Supply Chain Management at the I.H. Asper School of Business, University of Manitoba, Canada. He obtained his DPhil in transport geography from the University of Oxford. His research and teaching focus on port geography, management and governance, transport geography and regional development, climate change and the adaptation of transport and supply chain facilities, maritime logistics and global supply chains. He has published widely and has received a number of research awards and research fellowships. As a recipient of the Fulbright Scholar Program and the Endeavour Awards, he was a Visiting Scholar at Stanford University in the US and the Australian Maritime College and provided key strategic advice to intergovernmental and governmental agencies (including UN institutions) on various global transportation, logistics and supply chain issues. He is a Council Member of the International Association of Maritime Economists (IAME), as well as a Chartered Member of the Chartered Institute of Logistics and Transport (CILT) and a fellow of the Hong Kong Sea Transport and Logistics Association (HKSTLA). He is co-editor of the *Journal of Transport Literature* and a member of the Editorial Board of *Maritime Policy & Management* and the *Asian Journal of Shipping and Logistics*.

John J. Liu is Professor at the Centre for Transport, Trade and Financial Studies of the City University of Hong Kong, China. He has more than 20 years' experience of teaching, research and consultancy in risk and decision analysis in relation to logistics, supply chains and maritime services. After receiving his BSc and MSc in Marine Engineering from Huazhong University of Science & Technology, China, he continued his studies at Stanford University and Pennsylvania State University, where he obtained an MSc in Economic Engineering Systems and a PhD in Industrial Systems Engineering and Management, respectively. He has experience in leading and designing development programs for senior executives and professionals. He has been involved in a number of research projects and his articles have appeared in *Operations Research* and

Management Science. A recipient of a number of teaching and research awards, he has served on the Engineering Panel of the Research Grant Council, HKSAR (2005–2011), on the editorial board of *Maritime Policy & Management* (2008–2010) and as a member of the Human Resource Taskforce of the Maritime Industry Council, HKSAR (2006–2011).

List of Abbreviations

3PL	Third Party Logistics
ABDIB	Brazilian Association of Basic Industries and Infrastructure
ABEPRA	Brazilian Association of Dry Ports
ANTAQ	National Agency for Waterway Regulation, Federal Government of Brazil
ANTT	National Agency of Land Transport, Federal Government of Brazil
AP	Actual Productivity
APL	American President Line
APMG	A.P. Moller Group
ASEAN	Association of Southeast Asian Nations
BNP	Busan Newport
BOO	Build-Own-Operate
BOOT	Build-Own-Operate-Transfer
BOT	Build-Own-Transfer
BPA	Busan Port Authority
CBA	Cost Benefit Analysis
CCO	Chief Commercial Officer
CDS	Dock Company of Santos (*Companhia das Docas de Santos*)
CEO	Chief Executive Officer
CES	Constant Elasticity of Substitution
CFO	Chief Financial Officer
CFS	Container Freight Station
CIDCO	City Industrial Development Corporation, India
CLIA	Logistical and Industrial Centers (*Centro Logístico e Industrial Aduaneiro*), Brazil
CNI	National Confederation of Industries, Brazil
CONCOR	Container Corporation of India Ltd.
COO	Chief Operational Officer
COSCO	China Ocean Shipping (Group) Co.
CRM	Customer Relation Management
CRS	Constant Returns to Scale

CSX	CSX Intermodal Terminals, Inc.
CWC	Central Warehousing Corporation
DEA	Data Envelopment Analysis
DPW	Dubai Ports World
DTA	Custom Transit Declaration, Brazil
ERP	Enterprise Resource Planning
ERS	European Rail Shuttle
EU	European Union
EXP	Export
FDI	Foreign Direct Investment
FP	Frontier Productivity
GATT	General Agreement on Tariffs and Trade
GDP	Gross Domestic Product
GNP	Gross National Product
HHLA	Hamburger Hafen und Logistik A.G.
HPH	Hutchison Port Holdings
IBRALOG	Brazilian Institute of Logistics
ICD	Inland Container Depot
ICMS	Merchandize and Services Circulation Tax, Brazil
IDB	Inter-American Development Bank
ICTSI	International Container Terminal Service
IMP	Import
IT	Information Technology
JIC	Just-in-Case
JIT	Just-in-Time
JNPT	Jawaharlal Nehru Port Trust
KCTA	Korea Container Terminal Authority
KIPA	Korean Institute of Public Administration
KMI	Korean Maritime Institute
LPI	World Bank's Logistics Performance Index
MIU	Lloyd's Marine Intelligence Unit
MOMAF	Ministry of Maritime Affairs and Fisheries, National Government of South Korea
MPA	Maritime and Port Authority of Singapore
MRP	Material Requirements Planning
MSC	Mediterranean Shipping Co.
MTL	Modern Terminals Ltd.
MTO	Make-to-Order
MTS	Make-to-Stock

MVW	Ministry of Transport, Public Works and Water Management, National Government of the Netherlands
NEB	Non-Executive Board, Port of Rotterdam N.V.
NEG	New Economic Geography
NPC	National Port Council, the Netherlands
NPM	New Public Management (in the Netherlands)
NOL	Neptune Orient Line
NTT	New Trade Theory
NVOCC	Non Vessel Operating Common Carrier
NYK	Japan Mail Shipping Line (*Nippon Yūsen Kabushiki Kaisha*)
OGMO	Labor Management Entities (*Órgãos Gestores de Mão-de-Obra*), Brazil
OTM	Multimodal Transport Operator (*Operadora de Transporte Multimodal*), Brazil
OOCL	Orient Overseas Container Line
OSC	Ocean Shipping Consultants Ltd., UK
P&O	Peninsula & Oriental Steam Navigation Co.
PAC	Port Authority Corporation
PDDT VIVO	Master Plan for Transport Development (*Plano Diretor de Desenvolvimento dos Transportes*), State Government of Sao Paulo, Brazil
PE	Production Efficiency
POL	Port Operator Logistics
RDM	Rotterdamsche Droogdok Maatschappij
PoR	Port of Rotterdam N.V. (*Havenbedrijf Rotterdam N.V.*)
PRD	Pearl River Delta
PSA	Port of Singapore Authority
RMAFO	Regional Maritime Affairs and Fisheries Office, National Government of South Korea
RMPM	Rotterdam Municipal Port Management
RPA	Rotterdam Port Authority
SCM	Supply Chain Management
SECEX	Secretariat of Foreign Trade, Brazil
SISCOMEX	Integrated Foreign Trade System (*Sistema Integrado de Comércio Exterior*), Brazil
SFA	Stochastic Frontier Analysis
SSA	Stevedoring Service of America

TC	Total Cost
TCA	Transferred Channel Assembly
TCB	Terminal de Contenidors de Barcelona
TCE	Transaction Cost Economics
TCF	Transaction Cost Frontier
TEN-T	Trans-European Transport Network
TESCF	Transaction-Engaged Supply Chain Frontier
TEU	Twenty-Foot Equivalent Unit
TR	Total Revenue
UN	United Nations
UNCTAD	United Nations on Trade and Development
UNESCAP	United Nations Economic and Social Commission for Asia and the Pacific
UK	United Kingdom of Great Britain and Northern Ireland
US	United States of America
VAD	Value Added Distribution
WBD	Wideband Data
WTO	World Trade Organization
WWII	Second World War
YRD	Yangtze River Delta

1
Introduction

1.1 Setting the scene

Significant economic and policy changes have occurred within the global arena. The scale of the downturn resulting from the global economic crisis in 2008 is substantial, with its destructive force arguably even stronger and more widespread than that of the Asian financial crisis towards the end of the last century. It has not only affected the financial sector, but also international trade and the transport and logistics industries. Indeed, the maritime, transport and logistics sectors are closely knit with the well-being of the global economy – are indeed the artery of the global economy, carrying more than 80 percent of the world's cargoes (Ng and Liu, 2010) – and it is an opportune moment to investigate the way forward for these sectors, and to thrash out appropriate strategies. As noted by Liu (2009), such a challenge has prompted the industries to increase calls for resilient and disruption-robust port, logistics and supply chains. Transportation and logistics were traditionally separate, and segregated, economic activities; nowadays, following the inevitable trends of globalization, they have evolved into an integrated service profession. Nevertheless, mainstream supply chain research has hitherto been based on the theory of the firm as a production function. For instance, a supplier is often modeled as a function that generates specific outputs using certain input possibilities as independent variables. As production and supply chain operation are going global, two transactional characteristics can no longer be ignored, namely transaction costs and bi-direction production flows. Complementary to the model of the firm as a production function, a port-focal model allows a firm to be viewed as a governance structure, so as to attain organizational efficiency through, for example, the minimization of transaction costs through vertical integration.

Under this development direction, ports, being the nodal points, will play pivotal roles in determining the success and well-being of global supply chains; thus, the establishment of effective port-integrated logistical systems will become strategically important. How to achieve that, however, is a subject for discussion. For instance, towards the end of the last century, Slack (1999) proposed the establishment of 'satellite terminals' to allow ports to penetrate into inland areas. His idea was later elaborated by Notteboom and Rodrigue's (2005) 'port regionalization' concept. In this regard, dry ports, or 'inland terminals', are understood as inland settings with cargo-handling facilities, so as to facilitate various logistical functions, e.g. freight consolidation and distribution, temporary (in-transit) storage of cargoes, customs clearance, deferment of duty payment for imports stored in bonded warehouses, advance issue of bills of lading, transfer between different transport modes, relief of congestion in gateway ports, inventory management, and agglomeration of institutions (both private and public) within specific locations to facilitate interactions between stakeholders along the supply chains (Meersman et al., 2005). In this sense, they conduct many functions which complement port operations. There are alternative views that logistical, and indeed value-added, activities should remain within the port arena, conventionally termed 'port-centric logistics.' The idea is that cargoes (often containers) should be loaded and unloaded inside ports (or at very proximate locations) and the contents then carried inland in various forms (e.g. as palletized cargoes). According to Mangan et al. (2008), under this system, ports should not simply act as freight distribution trans-shipment points, but also play a pivotal role in catalyzing the establishment of efficient supply chains. In this regard, they offer benefits, but simultaneously pose considerable planning and management challenges, e.g. increases in cargo storage, stevedoring activities and congestion within port areas. On the other hand, the greater flexibility in distributing goods can lead to savings in the relatively costly inland transport section of supply chains. Hence, it is clear that port-centric logistics and dry ports/inland terminals are fundamental in enabling cargoes to enter or exit port areas more effectively and, as noted in an Inter-American Development Bank's (IDB) report, in improving hinterland transportation accessibility for both inbound and outbound supply chains (IDB, 2013).

Nevertheless, there are several issues on whether the adoption of port-centric logistics is sufficient to address the aforementioned challenges. The first questions are: under what circumstances will dry port/inland terminals and port-centric logistics work? For port-centric logistics, much of its ideas focus on imports, but how about exports? Simply put, this idea states that containers should be unpacked in the ports. However, in many developing countries, like India and Brazil, shippers mainly consist of small and medium-sized firms (see Chapters 7 and 8), and their cargoes may not fill even one container. Thus, the inland transport cost may simply become too high for them to move their (non-containerized) cargoes to ports. In this case, they need to rely on dry ports/inland terminals to insert their goods (together with other cargoes) into containers, and later transport them to ports, thus saving them costs. However, this involves the interactive dynamics between ports, cities and surrounding regions under diversified, and uncertain, environments in different parts of the world (Ducruet and Lugo, 2013; Fujita and Mori, 1996). In this regard, Ng and Gujar (2009a) provide a good illustration of the potential difficulties of the export of tea leaves from Northeast India to the international market without dry ports – and consequently of the competitiveness of that product. On the other hand, Monios and Wilmsmeier (2012) highlight the limitations of the use of port-centric logistics and dry ports in overcoming 'double peripherality' in relatively isolated regions like Scotland, whose challenges are largely due to current transport policy and settings, which limit Scotland's maritime access points to major English ports (e.g. Felixstowe, Southampton and Tilbury). In such cases, how can exporters reduce their costs?

The above leads to further questions: is port-centric logistics an alternative to dry port/inland terminals, or are they actually complementary and should they therefore work with each other? Dry port/inland terminal facilities will exist, but at the same time ports may construct new, and maybe largely similar, facilities. Will such a development direction lead to a waste of resources, and the risk of creating numerous duplicated facilities? If this happens, will these duplicated facilities (and thus ports and dry ports/inland terminals) compete fiercely with each other for cargoes and business? As illustrated in Chapter 8, this situation has already arisen in Brazil, and may become even more explicit if ports and dry ports are governed by different administrative authorities. In this case, can both ports and dry ports/inland terminals appreciate, and think from, a 'chain' perspective and develop hand in hand effective supply chains that can benefit all supply chain stakeholders? Or, on the contrary, will the bullwhip effect[1] be exaggerated, where stakeholders think only about their own business as an individual entity, and even treat their counterparts along the supply chain as competitors? In this regard, for countries with a huge landmass, like India and Brazil, going from inland production plants/markets to ports (and the other way round) often involves cross-border movements between states and provinces, and it is difficult to deny the possibility that coastal and inland states will compete with each other for resources and incomes. If this takes place, they will stimulate substantial political, institutional and administrative controversies. For instance, if port-centric logistics is fully put into practice, it means that all the goods will go to the ports and be put into containers. It also means that customs clearance will be done in the ports. Is this a phenomenon that inland (landlocked) states will be ready to accept (as this implies that they may lose their customs earnings)? As will be discussed in Chapter 7, in India, shippers often prefer to deal with 'local' customs officers (those based in the same state) so as to minimize uncertainties during the customs clearance process. Will they be ready to go to ports located in other states and deal with non-local customs officers?

All the above questions clearly indicate that, as supported by Majumdar (2012), port-centric logistics, and dry ports/inland terminals, may not be everybody's cup of tea, and that certain conditions must be achieved for dry ports/inland terminals and port-centric logistics to be implemented successfully. First, as mentioned earlier, both ports and other stakeholders along the supply chains need to think from a 'chain' perspective, where destructive inter-regional (e.g. between states/provinces) and inter-sectoral (e.g. between ports and dry ports) competition should be averted whenever possible. Also, there is a need to develop effective ways of minimizing, if not avoiding, the duplication of (capital-intensive) facilities and the application of different concepts in different countries and regions with diversified economic settings, given the fact that many emerging economies nowadays, notably China, India and Brazil, are export-led economies with ever-increasing domestic markets. Scholars, policy-makers, members of think tanks and industrial practitioners must find effective ways to solve these problems.

1.2 Objectives and contributions

This book addresses the trends and challenges that the maritime, transport and logistics industries have tackled in recent decades, and the way forward for the development of efficient logistics and supply chains. It provides scholars, policymakers, members of think tanks, industrial practitioners and students with a comprehensive view and understanding of a growing, but also rapidly changing, sector, especially in developing economies. When writing this book, the authors recognized the limitations of both the development of dry ports/inland terminals and port-centric logistics, and thus a new concept – port-focal logistics – will be proposed and its implications on ports, logistics and supply chains assessed. They appreciate the valuable efforts provided by previous researchers in the study of the development of port, logistics and supply chains, but simultaneously recognize their limitations – due largely to the research trends of logistics and supply chain management being dominated by an emphasis on optimization rooted in the operations management and management science disciplines – which imply the need for further improvements and fine-tuning of these concepts in a world of globalization.

To achieve this requires efficient linkages between knowledge in different time periods (past, present and future) and sectors (academic, governmental and industrial). This book reviews the fundamental elements of port and logistics management, as well as the impacts of contemporary shipping development on ports, logistics and supply chains around the world. In this regard, the authors strongly believe that ports are excellent sites to interrogate the wider applicability of institutional analysis in explaining the change and evolution of logistics and supply chains, especially given their fundamental transformation in the past few decades. This transformation has created substantial pressure to re-invent the nature and philosophy of port governance through diversified means. Specifically, the book focuses on how the governance of ports in different corners of the globe is embedded within the higher levels of political and institutional structures. In turn, it investigates how institutional legacies contribute to diversified outcomes in diversified economic, social and political contexts. Indeed, all subsequent chapters will circulate around, and provide sound justification for, the main idea and revelation that this book will bring to readers: the development of port-focal logistics is the appropriate direction for future global supply chains.

It is noted that much of this book focuses on developing economies located in Asia and Latin America. A substantial amount of unpublished, and much confidential, information was collected during a number of research field trips in the past several years, during which more than 70 semi-structured, in-depth interviews with relevant personnel within the port, logistics and supply chain sectors in different countries and regions was conducted. As a result, this book provides first-hand insight into what strategies developing economies should (or should not) adopt when developing their ports, logistics and supply chains under diversified, often uncertain, circumstances. Until now, there has been very little research (especially in English) specifically focusing on the development of ports, logistics and supply chains in these parts of the world. Also, by analyzing the issue from many different geographical, sectoral and disciplinary angles, the book will give readers not only an alternative view but, more importantly, the rare opportunity of an epistemological reflection on the true evolution and development of ports, logistics and supply chains in this ever-changing world, and will equip them with the wisdom and knowledge necessary to the development of effective solutions to the stated challenges.

1.3   Structure of this book

The rest of this chapter will introduce the structure of this book. Chapter 2 will start with a brief history of the evolution of shipping, notably the factors that have stimulated the development of shipping and its myriad manifestations. It will provide a brief history of shipping until the 1950s, followed by an account of the development of shipping since then. The impacts of this development on ports will then be illustrated. This will offer readers an important background to the major topic of this book, as contemporary international trade, and indeed the development of logistics and supply chains, is closely linked with international shipping. Simultaneously, as an inevitable result of globalization, supply chains and logistics have gone global. In the light of this, Chapter 3 reviews the mainstream supply chain and logistics studies, which are mainly firm-focal. After that, the chapter investigates the operations management of global supply chains, together with global outsourcing as an operational driver towards port-focal supply chains and trade logistics. This chapter also studies the operations management of port-focal supply chains and trade logistics themselves.

Indeed, containerization affects not only the development of shipping, but also other components of the transportation and logistical systems, including ports and inland transportation. Hence, Chapter 4 will focus on the interaction between shipping, ports, logistics and supply chain development. It will discuss how containerization, the restructuring of shipping and the development of global supply chains have shaped the evolution and development of ports. It will start by providing an assessment of ports as the key components in logistics facilitation. This will be followed by an explanation of how their development, together with the popularity of neoliberal ideology in economic policy development in recent decades, has prompted ports to undergo management and governance reforms, aim for hinterland access and establish logistics hubs. In this regard, the effectiveness of port-integrated logistical systems will surely depend on the productivity and efficiency of ports, other components of the supply chains and trade industrial organization, and Chapter 5 will address this issue. First, port and supply chain productivity and efficiency will be investigated, since they serve as key performance measures of port-focal supply chain management. However, compared with the firm-focal supply chain, the port-focal supply chain contains ports that are not only production facilities, but also transactional ones. Hence, it will continue to study the transaction cost economics of port, transport and international trade, pertinent to organizational dynamics of port-focal supply chain and logistics. At the end of the chapter, port efficiency and performance assessment will be investigated, with an empirical case relating to global performance benchmarking of container ports. In this exercise, the authors identify *environment heterogeneity*, notably policies and institutional factors, as a critical factor often overlooked by researchers when studying the evolution and development of ports, logistics and supply chains, duly supported by the experiences of governance in the ports of Singapore and Hong Kong. Indeed, this is especially true in developing economies, where the influence of institutional agents is usually very significant, if not pivotal.

This issue of *environment heterogeneity*, and its impacts on ports, will be further addressed in Chapter 6. It starts with a discussion on the evolution of port systems and continues with a detailed theoretical discussion on the roles of institutions in changing the nature, philosophy and implementation of economic activities. After that, it will provide two case studies on how institutions have affected the governance structure of two major ports in Asia and Europe, namely Busan (South Korea) and Rotterdam (Netherlands) by corporatizing their (initially public) port authorities. To elaborate on this topic, the influence of institutions on the development of ports, logistics and supply chains will be further discussed in Chapters 7 and 8, using case studies from two of the world's major developing economies, namely India and Brazil. They focus on how policies and the embedded institutional systems of both countries have restricted the integration of ports into their respective supply chains, and how ports and dry ports are unable to achieve their objectives of providing effective logistical functions to shippers and other logistical stakeholders. This issue has rarely been investigated in detail in developing economies, and the inclusion of experiences from two of the world's most important developing economies will greatly strengthen the value of this book. In these chapters, the authors will also question whether so-called Western solutions can be directly applied to developing economies: if so, how, and if not, why. At the end of the chapter, they will further investigate the relationship between institutions and the effective development of port-integrated logistical systems.

Chapter 9 will summarize the major findings, and will provide sound justification for the main idea and revelation that this book formulates: the development of port-focal logistics as the appropriate direction for future global supply chains. To this end, it will consist of a discussion on the pioneer observation on the 'repulsive' nature of the bullwhip effects on supply chains located in countries or regions with strong influences from environment heterogeneity, notably institutional factors, and how such repulsion may trigger the self-destruction of supply chains by supply chain stakeholders. In turn, it also highlights the inadequacies of existing theories on port evolution and development, as well as the continual emphasis on firm-focal supply chains by mainstream supply chain researchers.

1.4 Acknowledgment of previous publications

Some of the contents of this book exist in different forms published in scholarly and professional journals, conference proceedings and a doctoral thesis. They include Ng (2006), Ng and Pallis (2007a, 2007b, 2010), Ng and Gujar (2009a, 2009b), Yan et al. (2009), Ng and Tongzon (2010), Ng and Cetin (2012), Padilha and Ng (2012) and Ng et al. (2013).

2
Contemporary Development of Shipping and Impacts on Ports

The evolution and development of shipping reflect the human relationship with both the physical landscape and the built environment of the Earth, which is mostly covered by water. This involves the study of various disciplines, such as geography, economics, engineering and the other social sciences, and its recent development is pivotal in affecting the development of ports, logistics and global supply chains. For this reason, this chapter provides a brief history of the evolution of modern shipping, and endeavors to understand the factors stimulating the evolution and development of shipping and its myriad manifestations, and how shipping affects ports, logistics and supply chains. In doing so, it offers readers a fundamental background on the major topic of this book.

2.1 A brief history of shipping before the 1950s

Although the history of shipping is not the scope of this book, it is necessary to provide a brief account here so that readers can identify the substantial developments that the sector experienced before and after the 1950s. For further details of the history of international shipping, readers should refer to the book edited by Harlaftis et al. (2012).

There is ample archaeological evidence to suggest that seaworthy boats existed even during the pre-historic period, in the form of dugout canoes mainly used for coastal fishing and travel. However, it is generally agreed that shipping as an organized activity developed around 3000 BC, with navigation along rivers and coastal areas by a number of great ancient civilizations, in Babylon, Egypt, the Indus Valley, Phoenicia and Greece, to name but a few, mainly for fishing, trading and military purposes. Indeed, the Hellenic tradition of developing maritime power was later taken up by the Romans in their endeavor to build their empire; notable was its pivotal role in the overthrow of the Carthaginian Empire during the Punic Wars (264–146 BC).

During the Medieval era, merchant fleets of the major western European powers, notably Portugal, Spain, Great Britain and the Netherlands, circumnavigated the globe and established new trading routes in search of sources of raw materials like spices, gold and silver to feed the growing populace of Europe, partly encouraged by mercantilism, the fall of Constantinople (and thus the East Mediterranean) to the Ottoman Empire, and the new, adventurous ideas inspired during the Renaissance. During the process, adventurers successfully sailed around the southern tip of the African continent (Illustration 2.1) and, at the same time, discovered the hitherto unknown lands of America. The discovery, followed by the rapid colonization of the New World, resulted in fierce competition

Illustration 2.1   The Cape of Good Hope at the southern tip of the African continent, where the Atlantic and Indian Oceans meet

Source: Authors, taken at the Cape of Good Hope, South Africa (2010).

between the developing European nation states. During this period, shipping consolidated its position as the artery of international trade, wealth and, ultimately, political power. In the following centuries, it gradually extended the Europeans' maritime reach to Asia and Australia (Illustration 2.2).

Illustration 2.2   HMS Endeavour (replica) was the first ship to reach Australia, in 1770

Source: Authors, taken in Melbourne, Australia (2012).

This trend continued, and was strengthened, during the Industrial Revolution, which further boosted demand for raw materials from the colonies (such as American cotton). In turn, it demanded bigger, faster and more reliable ships. The technological gains of the Industrial Revolution were used for the development of steam-powered ships, which were not at the mercy of wind or tides and could enter and exit ports at will, which, in turn, permitted the introduction of scheduled shipping services (Williams and Armstrong, 2012). Also, they could carry large quantity of raw cotton to the textile mills in Great Britain, and resulted in making it a prosperous and wealthy country. This wealth was used by Great Britain (and later other European nations) for further improvements in technology, industrial development and colonial expansion, and gradually established the so-called staple-based economic system (Harley, 2012) between the European powers and their colonies. By 1900, there was already a high intensity of steamship services, both passenger and cargo, to most parts of the world (Williams and Armstrong, 2012). Ekberg et al. (2012) have even argued that shipping played a pivotal role in fostering a truly global economy before the Great War (1914–1918).

Given such political and economic significance, together with the increasingly capital-intensive nature of the industry thanks to the deployment of steam-powered steel ships, shipping gradually became a highly protected sector in many of the major European colonial countries. Indeed, it was during this period that liner shipping 'conferences' were introduced, such as the UK–Calcutta (1875) Conference and the UK–Far East Conference (1879). All conference members would charge a commonly agreed fixed freight rate to shippers. Moreover, members would discuss and agree on common rates for different commodities. To avoid competition among members (as well as from non-conference members), further actions were usually introduced. First was the pooling of capacity and revenues. In many cases, the amount that each shipping firm might carry was discussed and agreed upon, while extra revenues generated by one or more firm(s) needed to be shared with other members within the conference. Also, in order to dilute potential threats from external competitors, entry barriers were often imposed on non-conference members, notably in the form of predatory pricing and exclusive contracts for shippers. The consequence was the minimization of diversification, in terms of both services and quality, between carriers. On the positive side, it ensured the stability of shipping services and schedules for nearly a century. This was especially true of liner shipping, where a pre-advertised time schedule, dictating that a ship must leave the port regardless of whether it was full or not, ensured that its costs were fixed and independent of the types and amount of cargoes being carried.

Later, during the inter-war period (1918–1939), improvements in technology, especially in electrical power generation, transport and machinery, triggered a sharp demand for cheap oil, shipped especially from the Middle East to Europe via the Suez Canal. After the Second World War (WWII) (1939–1945), a byproduct of the mass production during the early postwar period was the utilization for the ease of handling and availing benefits of economies of scale. Consumerism coupled with mass production and utilization led to the birth of containerization in 1956, when the US truck owner Malcolm McLean introduced his idea of using standardized steel boxes to accelerate the loading and unloading of ships. In the following year, he converted a tanker, the *Ideal X*, into a container carrier. Such steel boxes, now generally known as containers and mostly standardized as Twenty Foot Equivalent Units (TEU),[1] brought shipping into a completely new era, which lasted until today.

2.2 Containerization and recent development

With continuous annual growth in international trade since the oil crisis in the 1970s, the shipping industry had grown rapidly (Peters, 2001). With shipping carrying more than 80 percent of internationally traded goods today (Ng and Liu, 2010), it is expected to continue its role as one of the most important components in global trade in the foreseeable future. Being a part of the global transport system, shipping creates economic benefits for globalization, enables the breeding of regional specialization (Hanson, 2000) by linking maritime corridors into networks and brings places closer together.

In the past five decades, a number of factors have shaped global shipping, including the growth of international trade, the emergence of new markets, the global division of labor, the regional specialization of production and the emergence of new carriers. This was encouraged by the popular trend of having a direct presence in particular overseas locations or markets, complemented by the successful conclusion of the Uruguay Round of the General Agreement on Tariffs and Trade (GATT) (1986–1994) regarding the removal of trade barriers. However, this process would probably not have taken place without the support of efficient transport connections in supporting the sophistication, productivity and thus the competitiveness of firms in the contemporary world (Porter, 2000). Hence, it was essential for transport operators to offer better access between manufacturers and consumers, allowing specialization and geographical concentration (Krugman, 1998). Towards the end of the last century, the global economy had entered a period where customers emphasized greater variety, customer-oriented business strategies, adaptation to individual demands and better product quality. As the tastes of customers changed frequently, goods should be delivered within a short, but precise, period, commonly known as a 'just-in-time' (JIT) strategy, which will be further discussed in Chapter 3.

These factors, complemented by technological innovations, contributed to huge capital investments and the mechanization of cargo handling. In this regard, the landscape of shipping changed dramatically, with increased demands for more reliable cargo movements, which reinforced the concept of multimodal transportation, logistics and global supply chains. As shipping was the artery of the global economy, it also led to the fundamental structure of maritime transportation, both ships and ports. As early as the 1970s, Bennathan and Walters (1979) foresaw that the world would witness the increasing importance of trans-shipment within the industry, with ports conceding their monopolistic position. They argued that competition between ports would be far from just a cost war, and that quality of service would become equally prominent.

Moreover, the new environment required a new operational and management philosophy. According to Notteboom and Winkelmans (2001), to succeed in this environment, suppliers and producers should possess flexible business attitudes and strategies to meet diversified demands. With continuous development in international trade and production, most shippers expected innovative transport systems for the smooth delivery of commodities, and had high expectations of transport operators, which included shipping companies. With such changes in perception, instead of merely spreading fixed costs to achieve economies of scale, transport operators needed to put more emphasis on customer taste, and to have wider geographical coverage and higher service quality.

Nowadays, container shipping is characterized by greater conformity, standardization, mechanization and homogeneity. As far as the ships themselves are concerned, the trend of substantial increases in ship size began in the 1980s (Chilcote, 1988) and continues today – and is likely to do so for the foreseeable future.[2] While the mean carrying capacity of the containerships of the top 20 shipping lines was 1,500 TEUs in 1989, this figure rose dramatically in the 1990s, with an average size of 2,000 TEUs and nearly 3,500 TEUs in 1994 and 1999, respectively. At the beginning of this century, the deployment of mega-containerships was a common practice among the top shipping lines (Illustration 2.3). For instance, CMA-CGM possessed a substantial number of ships with a capacity of 8,000 TEUs, while shipbuilder Hyundai Heavy Industries (Hyundai) had accepted orders for ships with a capacity of 13,000 TEUs (Hyundai, 2005). Towards the end of 2010, it was the norm for shippers to use containerships with a capacity of

Illustration 2.3   The use of containers has led to the construction and deployment of mega-sized ships with large capacity

Source: Authors, taken in Rotterdam, The Netherlands (2010).

more than 8,000 TEUs on the busiest shipping routes (notably Far East–Europe and Far East–North America), while there are currently plans to construct containerships with a capacity of up to 20,000 TEUs. Although initially some researchers questioned whether such a trend was sustainable – like Johnson and Garnett (1971) and Bekemans and Beckwith (1996), who worried about the limitations of port infra- and superstructures, the implications of the oil crisis and (increasingly important) environmental concerns – developments since the 1990s, and until today, have suggested otherwise.[3]

The continual increase in ship size supported the view of Slack et al. (2002), who argued that more uniformity and concentration was emerging within the shipping sector. Due to the inevitable trend of globalization and to keep pace with the development of other sectors of the global economy, shipping lines sought to further rationalize its structure through increasing their services. For instance, container services provided by the top 20 container shipping lines rose from just over 400 in the late 1980s to nearly 600 in the late 1990s. Some scholars went further and argued that the implications had shifted from pull to supply-led strategies, where cargoes followed ships (for instance, Chilcote, 1988; Cullinane and Khanna, 2000). With the economies of scale due to ship size and increasingly easy accessibility to advanced technologies, shipping lines managed to adjust development strategies based on their own perceptions. There were two main reasons for this phenomenon. First, the anarchic nature of the open sea had led to the lack of artificial forces in affecting its uniform pattern.[4] Such a deregulated environment enabled shipping lines to switch network operations for purely cost reasons and the environment operated under free market conditions (McCalla, 1999). Whereas public authorities could often impose licenses and regulations on other transport modes, they had limited influence on (deep sea) maritime transportation – not helped by the popular use of the flags of convenience in the past few decades. Second, technological changes required more capital investment in dedicated facilities for the handling of cargoes. Hence, firms possessing the financial muscle in offering quality service gradually decreased, or requiring a combination of strength. As noted by Palmer (1999), containerization occurred so rapidly that the shipping sector was not even allowed an evolutionary response. With continuing increases in ship size and services, nearly all the major shipping lines perceived that they could survive only if a scale economy approach was included, and effectively adopted, in their development strategies.

Given the highly speculative nature of investments, the economy of scale approach in shipping seemed necessary and it had gradually developed into an expensive lottery with only those who were willing to take the high risk of heavy investment having any chance of winning, leaving those who were unable to participate in the introduction of larger ships doomed to near-certain failure. Speculation in global shipping had completely altered its landscape; despite its high cost, containerization occurred at a surprising rate from the 1970s not only because of confidence in containers, but also, if not more, out of the fear of destructive competition. This meant that shipping investments must preserve the delicate balance between providing capacity to meet customer demand and creating overcapacity. As Notteboom (2002) argued, a scale economy approach in a capital-intensive industry like shipping could lower market contestability due to the significant amount of sunk costs, which substantially reduced the ease of market entry and exit. Thus, once commitments on new equipment and services had been made, considerable obstacles existed for investors to abandon them, even if it were clear afterwards that demands for new services had been overestimated and that profit gains were minimal. With such a dilemma, the only way for them was to move on and try to get as much as possible. Furthermore, the increase in ships' size was directly proportional to the financial cost of their construction, i.e. the high proportion of fixed costs in the total cost relative to the cost of maintenance and operation. Rising prices seemed to achieve little in encouraging shipping lines to drop their investment plans, especially the major ones. As Haralambides (2005) put it, many shipping lines chose to continue purchasing or chartering mega-containerships even if they could foresee that they would not be able to fill them.

With high fixed costs, only a small part of the total costs could be avoided even if the ships were left idle. Such growth created significant challenges because shipping lines needed to add capacities to their fleets within a very short period (Slack, 2004). This resulted in the higher opportunity costs of larger ships waiting in ports than smaller ones. Also, larger ships created further challenges, notably higher insurance costs, the difficulty of serving secondary routes and the more complicated loading and unloading system, and thus potentially more idle time (Stubbs et al., 1984; Chilcote, 1988). This meant that the cargo demands of the more isolated, peripheral regions were not enough to fill these ships. From the perspective of many shipping lines, these regions should be 'given up,' and their mega-ships should concentrate on the trunk routes, where there would be adequate demand to fill their ships (and slots). Despite the anticipated market demand increases, these did not necessarily keep pace with the increasing size of the ships. Even after several decades, the unfavorable situation of liner shipping, characterized by its capital-intensive nature but relatively low revenue returns (Chrzanowski, 1975)[5] had not really improved. In some cases, it even became worse. It implied that the competition within the shipping sector had intensified at a somewhat unexpected rate (Slack et al., 1996).

Thus, it was necessary for shipping lines to ensure that the exploitation of economies of scale would pay dividends, implying that their ships must be fully (or mostly) loaded and in operation more frequently (Liu, 1995). Such requirements would pose serious challenges to ports. As ships needed to operate more frequently, fewer ports should be used. As Slack (1998) and Junior et al. (2003) pointed out, the evolution of shipping made higher demands on port efficiency due to the need to maximize the number of voyages and minimize port stays for mega-ships. In short, ports would face immense pressure to enhance their services, while at the same time face the risk of leaving new equipment idle because of the failure to attract users.

At the same time, it was clear that development in the 'hard' aspects needed to be supplemented by the 'soft' components. With intensified competition between shipping lines, they needed to widen their geographical coverage so that demand could sustain the increased supplies. Simultaneously, they needed to ensure that calling at fewer ports would not affect the geographical coverage of the shipping networks. This dilemma led to various restructuring programs within the shipping sector. First, shipping lines gradually established the 'trunk-and-feeder' system, where a few 'best' ports within particular regions would be chosen and inter-continental ships would stevedore there, while smaller, short sea ships would carry cargoes, notably containers, to the final destinations. An illustration of the transformation from multi-port-of-call to trunk-and-feeder systems can be found in Figures 2.1a and 2.1b.

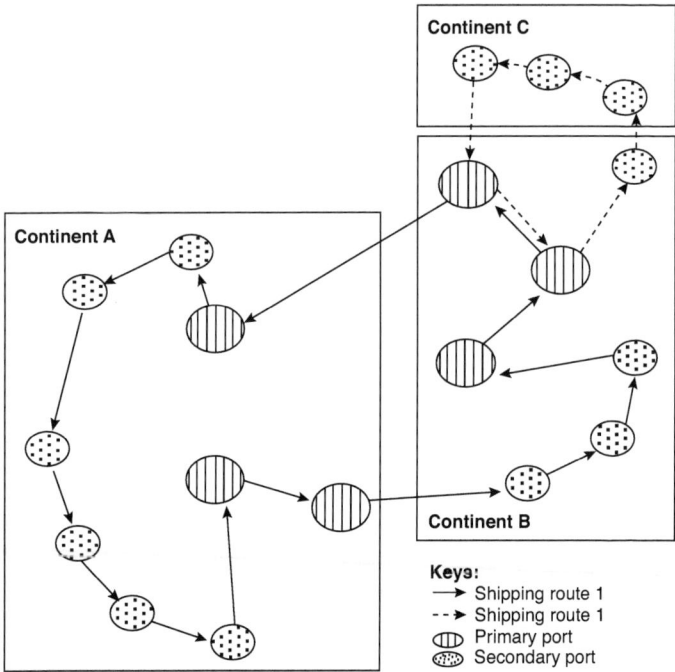

Figure 2.1a Multi-port-of-call shipping network before containerization
Source: Authors.

Under this system, the best ports become transshipment hubs where containers would agglomerate and be distributed throughout the region, while shipping lines could serve more geographically isolated markets, which initially did not economically justify direct linkage. However, as mentioned earlier, this meant that only the few chosen ports could benefit, as the current business of other ports would likely be taken away with cargoes being agglomerated geographically to a few hubs, especially since mega-ships implied the need for wider access channels and deeper water. The result was the establishment of a hierarchy of ports and a system whereby regional or local hubs would be connected by feeder vessels to outlying ports (World Bank, 2007). Large ships would be utilized to provide services between regional hubs and smaller ships would be used to pick up and distribute containers within the region. In their recent work, Wang

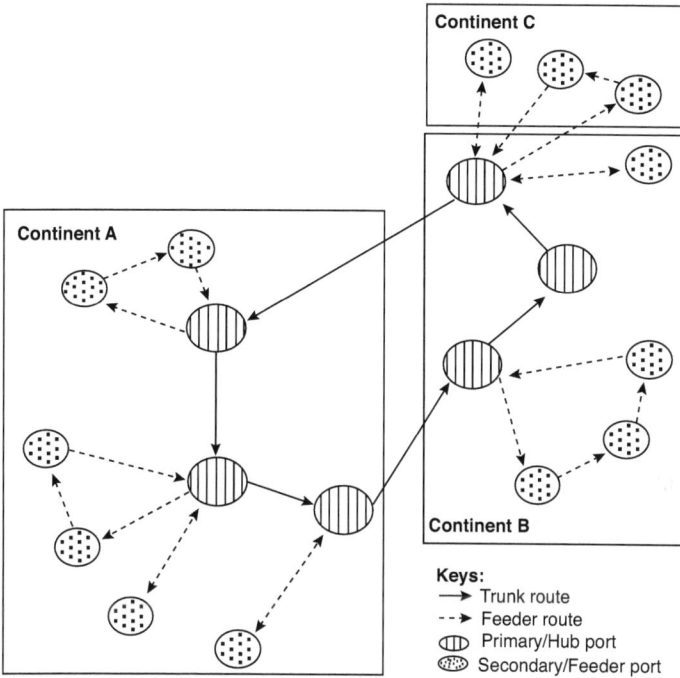

Figure 2.1b   Trunk-and-feeder shipping network
Source: Authors.

and Ng (2011) have identified at least three levels of hierarchies within the Chinese port system, providing different levels of container shipping service between Chinese ports and regions around the world.

Further reinforcing the scale of operations, shipping underwent another restructuring, involving acquisitions, mergers and the establishment of strategic shipping alliances, such as the takeover of American President Line (APL) by Neptune Orient Line (NOL), the merger between the Peninsula & Oriental Steam Navigation Co. (P&O) and Nedlloyd (P&O Nedlloyd eventually being acquired by Maersk) and the formation of the Global New World and CKYH Alliances and more recently, the proposed P3 Alliance between Maersk, CMA CGM and Mediterranean Shipping Co. (MSC). According to Kumar (2000) and Heaver (2002), the main reasons for horizontal integration included the protection of global market shares, reduction of costs through economies of scale, better perceptions of the market through information exchange, chances to enter new markets and stronger bargaining position vis-a-vis ports. Simply speaking, it evolved around the need for greater control in the decision-making process, a combination of financial power to expand and the sharing of financial risks. Apart from financial reasons, horizontal integration helped enhance the shipping networks' geographical coverage. In this sense, the restructuring of liner shipping was not dissimilar to the neo-liberalization of other economic policies, where, towards the end of the last century, dramatic increases in horizontal integration had also been found (Dunning, 2000). Acquisitions, mergers and shipping alliances focused on operational cooperation such as re-scheduling liner routes, which not only relieved competition, but also enhanced the overall competitiveness of stakeholders within the shipping sector.

On the other hand, the formation of strategic shipping alliances enhanced the market presence and geographical coverage of the shipping lines, and allowed them to construct ships with even higher capacities, as individual firms would now need only to concentrate on the operation of certain shipping routes that they excelled in. In other words, the formation of strategic shipping alliances was to achieve full ownership and control within a corporate umbrella (World Bank, 2007). This process became so rapid that, by early 2014, nearly all the top 20 shipping lines had joined one or more strategic shipping alliances (and even some arrangements between different alliances). As the sector's development matured, completely individual shipping lines had become the exception, while large, allied firms gradually became the norm. Shipping lines seemed to believe that they had found the benefits combining collaboration and competition in their business strategies. This complemented the view of Slack (2004), who argued that the force of globalization had finally led to rapid restructuring of the geography and structure of shipping.

2.3 The impacts of the recent development of shipping on ports

Network extensions could lead to lower cost of shipping services and greater geographical coverage. However, it also exerted upward pressure on both monetary and non-monetary costs (Bergantino and

Veenstra, 2002), especially the risk of missing feeders due to p
ficiency. As noted by Notteboom and Winkelmans (2001), s
shipping alliances were unstable. To consolidate alliance stability,
members tended to impose rather strict contracts and agreements
among themselves. In this sense, a vicious cycle was established,
as strong binding could cause problems among shipping lines and
increase costs of their operations since the decisions and strategies of
liners could become rather complex.

Such an underlying problem, together with the generally accepted
view that dramatic rises in freight rates to shippers were very difficult, if not impossible (Panayides and Cullinane, 2002), implied that
the minimization of costs was an important objective that shipping
lines must achieve. Since economies of scale in shipping were countered by increasing diseconomies of scale in terms of logistical costs
(Haralambides, 2000), ports became the critical components in determining such success. Thus, from the 1990s, shipping lines tended
to pressurize ports to improve, as the benefits of the economies of
scale and alliances would not be realized unless their spending on
ports could be dramatically reduced. Here, costs did not only involve
monetary ones, but also other, less readily quantifiable, costs, like
efficiency and quality of service offered by the ports. As argued by
Huybrechts et al. (2002), ports which contributed most in substantially reducing the generalized cost of transport chains would be
the most likely to be selected. As stated by van Klink and van den
Berg (1998), in contemporary liner shipping, direct monetary costs
did not alone necessarily determine the competitiveness of a port
because port choice involved other non-monetary factors.

As mentioned earlier, the above issue was raised as early as the
1970s, when there was concern that many ports would not be able
to handle a large number of containers and thus potential increases
in throughput, while the reduction of port costs was difficult. In
this regard, Chilcote (1988), Notteboom (2002) and Panayides and
Cullinane (2002) noted that large ships could not waste time and
call at additional ports for more cargoes due to high capital costs and
port costs in serving them. Since the development of large, higher-
speed ships was deemed economically impractical in the foreseeable
future (World Bank, 2007), the service frequency and turnaround
time of large containerships had to increase and decrease, respectively, while the number of port calls, as mentioned earlier, must

be reduced. As Chilcote (1988) pointed out, before containerization, ships usually called at a dozen ports within a continent within a liner service, compared with a maximum of five since then (Figures 2.1a and 2.1b).

Another implication of the contemporary technological and economic development of shipping was that it had significantly enhanced the bargaining power of port service users vis-a-vis service suppliers. With improved transport networks through technological innovation, shipping had become more footloose in port selection, and thus the traditional understanding that ports possessed certain natural hinterlands and thus few worries about losing customers had become obsolete. In this regard, McCalla (1999) argued that ports were increasingly becoming the 'servants' of their shipping lines, while Hayuth and Hilling (1992) noted that, as the world moved towards more borderless trading, discrete natural hinterlands would break down and be replaced by 'common hinterlands.' This factor had considerable implications in terms of inter-port competition. Shipping lines gained immense powers in controlling the fate of different ports, especially since ports had increasingly become part of the maritime logistics and supply chains (Heaver, 2002) (to be further discussed in Chapters 3 and 4). Also, Slack (2004) argued that the radical changes in ports of call had contributed to the growing similarities between shipping services and led to the situation where the winners would win more and the losers would lose even more. Indeed, until recently, when conducting their business, ports usually negotiated with their customers independently, and thus shipping's bargaining power was enhanced through better information, which allowed them to play the same threatening game with different ports.

Thus, the competition between major ports has intensified in the past decades. With changing demands, the traditional philosophy in port management, characterized by strong public presence, seemed no longer valid – a notion that was further strengthened by the increasing influence of neoliberal ideology in the development of economic policies during the same period (see Harvey, 2005). The intensified inter-port competition was especially significant in Europe, where the emergence of a single market had exerted pressure on national and regional authorities to give up protective measures regarding their respective ports, notably state aid (Hinz, 1996). Not to be left out, many ports continuously improved facilities and services and tried to attract new users (Meersman and van de Voorde, 1998). As Stone (1998) and McCalla (1999) pointed out, the market shares of ports were no longer guaranteed and the global actions of shipping lines often led to port sufferings. This trend spread to the major economic powerhouses in the rest of the world at the turn of the century, the Pearl River Delta (PRD) in Southern China (cf. Homosombat et al., forthcoming; Lam et al., 2013) being an example.

Facing such pressure, while shipping lines invested heavily to accommodate its own development, the commitments from ports had become even greater. According to Slack (1998), improvements in port infrastructure and services, usually accompanied by significant capital investments, were required so as to reduce ships' waiting time in ports, and smoother documentary procedures were needed so as to maintain their competitiveness as they transformed themselves into load centers (Chilcote, 1988). Ports needed to be specially designed, and be more innovative, so as to add value to their services and be more competitive in today's highly competitive environment. This exerted substantial financial pressure on many ports. Also, such new development required high load factors, and thus demands for quality services on container ports had become more demanding than ever (Slack, 1998). All these forces had pushed ports to become more complex and multifunctional (Robinson, 2002). Indeed, it was not long after the container revolution that the labor-intensive landscape of ports was replaced by a landscape dominated by expensive infrastructures and superstructures (like cranes and container-moving vehicles) (Illustration 2.4). At the same time, many ports had relocated from city centers to more remote locations due to increased space requirements and high land prices (Campbell, 1993). An example was the port of Rotterdam, originally in the city centre; its latest expanded facilities, *Maasvlakte 2*, are located more than 50 km away (Maasvlakte 2).

Of course, even after such expensive investment, there was no guarantee that ships would come. Haralambides (2000) warned that port development, notably physical infrastructure, was often built far ahead of existing demands owing to the perception that they would

Illustration 2.4   The labor-intensive landscape of ports has been replaced by capital-intensive cargo-handling facilities since containerization
Source: Authors, taken in Shanghai, China (2008).

lose out by not doing so. In this regard, the PRD serves as an example, where many container terminals have been constructed within an extremely small area in the past two decades (see Wang et al., 2012). His concern was shared by Psaraftis (1998) and van Ham (1998), who warned of the risk of ports being trapped in a vicious circle of creating wasteful overcapacity due to panic and unnecessary speculation. Similarly, Powell (2001) emphasized that any long-term plans could pay dividends only if the resources spent on infrastructures did not exceed the benefits derived from those particular investments. As Notteboom (2002) pointed out, the loyalty of port service users could not be taken for granted: not only because of potential deficiencies in port service, but, as mentioned earlier, also due to the restructuring of the shipping networks, new partnerships and strategic shipping alliances. McCalla (1999) put it even more strongly and claimed that the major challenge for ports was to find way to turn 'local pain' (like heavy infrastructure investments) to 'local gain' based on pure perceptions of the long-term demands of port services. Port operation had become an expensive lottery and investments in port facilities, usually durable with high sunk costs with significant risks (Hayuth and Hilling, 1992). Due to the decrease in natural hinterlands and the increase in the bargaining power of shipping lines, the pressure from increasing competition affected not only small ports but also large and established ones (Slack et al., 1996).

Thus, it became important for port managers to carefully evaluate their development plans, with needless duplication of facilities minimized (Rimmer, 1998). Indeed, recent developments within the shipping sector added substantial pressure to ports to operate with thinner profit margins due to intensified competition and heavy capital investments (Psaraftis, 1998). To enhance competitiveness, port managers were forced to alter their focus, from the traditional question of whether ports possessed the ability to handle a certain amount of cargo effectively, to whether they could attract potential customers, usually on the expenses of their counterparts. This posed significant implications for the philosophy in assessing port performance because the issue of 'port choice' had been put under serious consideration (Yeo et al., forthcoming). As long as port users were offered options to choose from, rather than measuring physical outputs, port performance within a competitive market should be evaluated on its ability to attract potential customers and to fight off competitors. To sustain competitiveness, ports must ensure that the increase in operation cost of shipping lines did not lead to significant increases in the value of the cargoes being carried (Gubbins, 1988).

In theory, sensible investments could be achieved through objective demand forecasting. As Peters (2001) had argued, objectivity was vital to avoid overstating or understating future container capacity and demand so that the ports concerned did not have to tackle serious congestion, nor waste substantial amounts of financial resources and leave expensive equipment idle. Nevertheless, the reliability and objectivity of these assessment tools were often doubtful. Forecast figures were sometimes exaggerated, a problem not helped by the increases in common hinterlands and trans-shipment traffic, the integration of port services into logistics and supply chains, and 'hub hopping' (World Bank, 2007). For instance, in the Hamburg–Le Havre port range in Europe, there were many established ports which could offer trans-shipment services and thus offer substantial choices to potential customers. A similar situation existed in the PRD in Southern China, where, as mentioned earlier, substantial (arguably excessive) investments in container terminal facilities had (unnecessarily) intensified competition between Hong Kong, Shenzhen and other PRD ports (Wang et al., 2012).

Also, the above indicated that port projects had become highly risky betting games and port managers must offer more capacity and better quality services (which might attract more business) without simultaneously creating overcapacity in relation to future market demands. Slack (1985) argued that rather than just monetary cost, quality of service would be an important attribute in deciding port performance and competitiveness. On the other hand, Xiao et al. (2012) pointed out that the ownership structure and other institutional arrangements of ports would also play a significant role. In fact, this had been recognized as early as the 1970s. Using competition between Singapore and Hong Kong as an example, Bennathan and Walters (1979) demonstrated that, due to increases in port size and investment in port development, the proportion of fixed costs became more important and inhibited the effectiveness of cost leadership, which in turn implied greater importance in differentials. As is to be discussed in Chapter 5, despite their similarities in many aspects, the ports of Singapore and Hong Kong have developed very different competitive strategies and governance structures.

To tackle these challenges, ports have to strengthen their own power against shipping lines and their counterparts, notably through the transformation of port management and governance towards a more business-friendly approach (as mentioned, this was facilitated by the endorsement of the neoliberal ideology in economic polices since the 1980s), and ports needed to be more responsive to users' needs than before (Juhel, 2001). Such requirements partly explained the increasing privatization of port operations around the world, in both developed and developing economies, and various research studies underlined this evolution (for instance, Baird, 2002; Heaver, 2002; Wang et al., 2004; Ng and Pallis 2010). For instance, under the 'landlord port' concept (World Bank, 2007), cargo loading/unloading in many ports was undertaken by private terminal operators. In this regard, Hutchison Port Holdings (HPH), the Port of Singapore Corporation (PSA) and Dubai Ports World (DPW) serve as examples. More recently, some port authorities have even transformed themselves into public corporations so as to allow more commercial activities and strategies to develop within the institution, with Busan, Piraeus and Rotterdam being notable examples (Ng and Pallis, 2010). The transformation in port governance will be further discussed in Chapter 6.

Perhaps more importantly, as mentioned earlier, recent developments in shipping had led to the growing importance of trans-shipment traffic, supplemented by the increasing global trade which greatly favored ocean-crossing shipping routes (Stubbs et al., 1984; Palmer, 1999). According to Taaffe et al. (1996), while the major reason for trans-shipment between different transport modes was the inaccessibility of maritime transportation to inland regions, trans-shipment within shipping was due to structural changes within the sector due to the aforementioned development. Although it could be dated back to the 1960s, when containers were first introduced (de Langen, 1998), trans-shipment did not accelerate until the 1990s, and it was actually a complement of the development of competitive hub ports. In the early 1980s, trans-shipment constituted only about 12 percent of global container throughputs (Damas, 2001), compared with almost 15 percent in 1990 and more than 20 percent towards the end of the twentieth century. Between 1990 and 2001, total container trans-shipment demand in Northern Europe increased threefold, from 2.18 to 6.72 million TEUs, and from 14.1 percent to 21.4 percent,[6] with an average annual growth rate of 10.86 percent. Indeed, with intensified competition for hinterland cargoes, some ports nowadays increasingly rely on transshipment to maintain their throughput growth and competitiveness, the port of Hong Kong being an example in the past decade (Wang et al., 2012).

The result of increasing transshipment was that major ports took dual responsibilities: keeping their traditional function of serving their respective hinterlands while adding the transit function that enabled them to become components of global, and more complex, logistical networks. Nevertheless, although the rise of trans-shipment traffic implied extra business for ports, this could be taken away by their users, i.e. shipping lines, at any time they deemed relevant, especially any innovations in reducing transshipment inefficiencies, notably the extra costs generated from the transfer of containers from one ship to another, were important to the competitive positions of liners themselves (Taaffe et al., 1996). In more developed regions, where ports were geographically proximate to each other, the rise of trans-shipment traffic also meant more intensified competition in the quality of services provided by ports (Hayuth and Fleming, 1994), the Hamburg–Le Havre range and the PRD being examples.

Clearly, the trunk-and-feeder system and increase in trans-shipment diminished the monopolistic nature of ports, while inter-port competition within the same geographical range intensified (de Lombaerde and Verbeke, 1989) as cargo flows became even more geographically concentrated (Heaver, 2002). Also, increases in trans-shipment traffic highlighted the problems related to port infrastructure, and ports were clearly the victims of such development, driven by shipping network restructuring, and had few opportunities to participate in this process (Ocean Shipping Consultants, 2004). Until quite recently, for example, limitations on water depth and locks along the River Scheldt partly explained why Antwerp focused more on trans-Atlantic than Far East–European shipping routes (the ships serving the latter routes usually being much larger). Furthermore, as shipping lines needed to avoid benefits being extracted due to price discrimination from ports, as mentioned earlier, they formed mergers, acquisitions and strategic shipping alliances to counteract the former monopolies of ports. The contemporary development of shipping suggested that the practice of price discrimination between transshipment and feeder traffic that had been commonly imposed by ports in the 1970s (Bennathan and Walters, 1979), of which the former usually enjoyed a much lower charge, no longer worked, even in the 1990s.

Although the rise of transshipment traffic could possibly increase overall port demands (Notteboom, 2002; Ha, 2003; Baird, 2004), as mentioned earlier, trans-shipment cargoes could make port competition even more intensified, as ships would be less likely to be restricted to particular ports. According to Hayuth and Fleming (1994), in a deregulated environment, an increase in transshipment traffic implied the rising prominence of 'intermediacy' (rather than 'centrality') in determining port competitiveness.[7] The increase in transshipment traffic and the need for effective strategic investment implied that ports needed to evaluate their respective positions, of which in turn would act as the platforms in establishing their future business strategies and success. Such understanding involved both exogenous and endogenous factors, including the foreseeable trend within the port and shipping industries, customer satisfaction with their services, development prospects and relations with their potential competitors. Moreover, through self-understanding, ports should develop strategies which could enhance their bargaining power over their customers so as to further enhance their competitiveness. Based on the above, Palmer (1999) was right to claim that revolution, rather than evolution, had taken place within the geography and economics of ports.

All the above factors had forced ports to move away from simple sea–land interfaces toward something more complex and comprehensive. In the past decade, many major ports have improved their connections with their hinterlands, as well as inland transport and logistics infrastructures. Simultaneously, within the port arenas, rather than just cargo stevedoring, many have started to provide more diversified, value-adding services and have targeted users from industrial sectors rather than just shipping lines. Nam and Song (2011) described this process as making ports into maritime logistics hubs, while some other scholars termed it port-centric logistics (see Chapter 1). Indeed, contemporary development of shipping had contributed very significantly to the transformation of ports, which will be discussed and analyzed in much more detail in subsequent chapters. At the same time, as an inevitable result of globalization, logistics and supply chains have gone global. Ports are playing more significant roles in this process. The next chapter will discuss the development of global supply chains and trade logistics, notably with regard to how they have evolved from being firm-focal to being port-focal.

3
Global Supply Chains and Trade Logistics: From Firm-Focal to Port-Focal

This chapter reviews the mainstream supply chain and logistics studies, which are mainly firm-focal. After that, it will investigate the operations management of global supply chains, together with global outsourcing as an operational driver towards port-focal supply chain and trade logistics. Finally, it will study the operations management of port-focal supply chain and trade logistics, the concept of *environment heterogeneity* being identified as an important missing factor in the development of ports, logistics and supply chains.

3.1 Firm-focal supply chain management

Mainstream supply chain and logistics studies have centered on the firm or enterprise, with the manufacturing firm as the primary prototype. A basic firm-focal supply chain can be considered as a three-echelon supply-order-production system, comprising inbound, factory/firm, and outbound stages of order fulfillment (Figure 3.1). The operations in each of the three stages are functionally differentiated, as follows:

Inbound stage: Purchasing, sourcing and outsourcing, and supply/service acquisition. Regular measures of inbound logistics include pricing, on-time supply and quality of supply/service.

Factory/Firm stage: Enterprise Resource Planning (ERP), especially capacity and inventory management, production planning and control. Regular measures of the factory stage are production cost, time-to-build, and product/process quality.

Outbound stage: Distribution, delivery and transportation. The regular measures are concerned with distribution cost, on-time delivery and customer service.

In normal cases, the transactions among the three stages are engaged sequentially (Figure 3.1). However, there are certain situations where some stages are bypassed by third-party logistics (3PL), either combined (through vertical integration) or networked (through strategic alliance). Specific cases of 3PL have been discussed in detail by Simchi-Levi et al. (2000). With advances in information technology (IT), innovative practices, like transferred channel assembly (TCA)

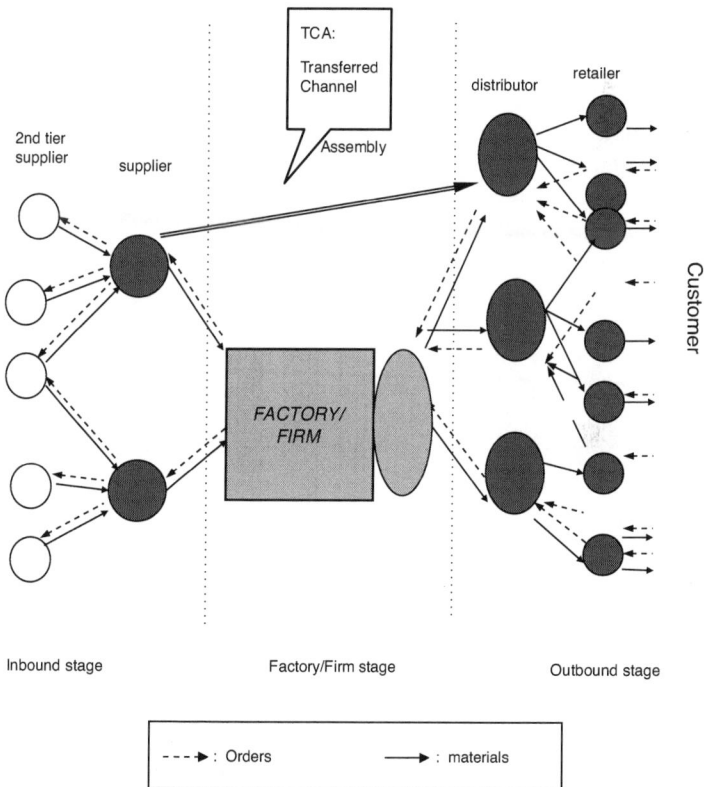

Figure 3.1 Simple supply chain with one focal factory
Source: Liu (2011).

and value-added distribution (VAD), have become more common in supply chain management (SCM). Here one should note that the interface and transaction between stages are facilitated *via* SCM, which unavoidably incurs certain transaction costs. In other words, the final market demand (in the future) is divided (in advance) into numerous supply contracts between various pairs of buyers and suppliers along the whole supply chains. A firm may play different roles in different contracts, as a buyer in one contract but as a supplier (i.e. a seller) in another. In sum, transaction cost economics (TCE), as coined by Williamson (2002), plays a vital role in logistics and SCM, as will be further discussed in Chapter 5.

3.1.1   Three flows in firm-focal supply chains

Although in reality a supply chain may be more complex, with more than three stages, all the stages of a supply chain are connected by 'flows,' which can be identified in three forms, namely information flow, material flow and financial flow.

Information flow refers to informational interactions between supply chain stages, such as information regarding orders and requests for needed supplies and products.

Material flow is the flow of delivered supplies and products in response to orders and requests.

Financial flow includes all the financial transactions associated with supply chain operations, which represent the core of the economics of supply chains.

From the perspective of firm-focal supply chain flows, a factory can be symbolically represented by a specific ERP system engaged with an inbound and outbound stage. Here it is interesting to note that an ERP system, by nature, is incompatible among factories, especially among factories which are competing with each other, while an inbound or outbound chain is, by nature, meant to be shared, and thus a real world supply chain is usually a partitioned flow network (Figure 3.2).

In Figure 3.2, the factories are competing firms, and each competitive firm must maintain an independent and firewall-protected ERP system. It is a representation of regular means and common technology available to competing firms. As a consequence, the 'winning secret' for a firm must be something other than ERP technology itself. In fact, such a secret lies in how a firm can uniquely operate and

Inbound chain		Outbound chain

• Material/Supply
• Purchasing
• Supply-based
  outsourcing (TCA)
• OEM
• MRO supply
• Technology
  acquisition
• Service contract
• Strategic alliance
	Firm 1
	(ERP-1)
	
	Firm 2
	(ERP-2)
	
	Firm *n*
	○
	(ERP-*n*)
	○
		• Warehousing
		• Transportation
		• Distribution and
		  delivery
		• Distribution-based
		  outsourcing (TCA)
		• Customer and post-
		  sales service
		• Recall and field
		  service

OEM: original equipment manufacturer. MRO: maintenance, repair, and operation.

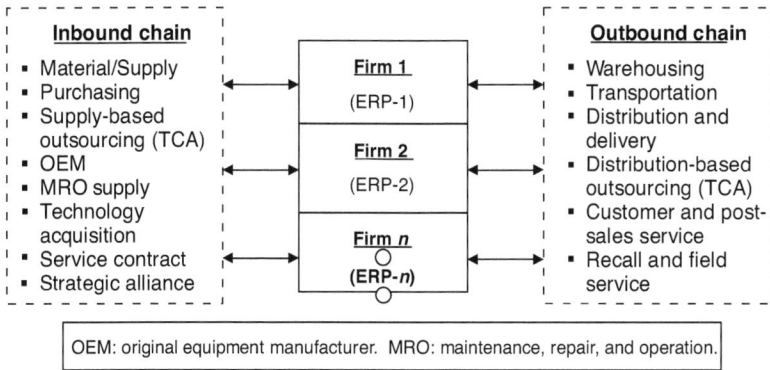

Figure 3.2 Complicated supply chain: a chain of competitive selling and buying
Source: Authors.

compete, given the same set of viable techn ologies that are publicly available. That is why not all firms using the same ERP system have similar success. Moreover, one should observe that the ERP systems in a supply chain are not suitable for sharing with other competing firms, thou gh the inbound and outbound resources are, by definition, to be shared by firms competing with each other.

3.1.2   Logistics vs. SCM

Logistics is referred to as the provision of supplies. This includes not only shipping and the transportation of supplies, but also relevant services and supports, e.g. warehousing (Illustration 3.1), customs clearance (Illustration 3.2), maintenance and insurance. In terms of professional contents and academic terminology, shipping encompasses all transport modes, while logistical services include all aspects of service industries, including finance, insurance, technology and infrastructures. Although a closely related discipline, SCM has been so far referred to by both scholars and industrial practitioners as the management of supply chains within a business organization which involve the provision of products and services to the end customers.

As evidenced by many textbooks, SCM has so far been production-based, typically in terms of manufacturing and service operations

Illustration 3.1   A warehouse in a dry port in India
Source: Authors, taken in Delhi, India (2009).

Illustration 3.2   A customs clearance station in a dry port in India
Source: Authors, taken in Delhi, India (2009).

management. Indeed, associated with a production-based supply chain, logistical activities, like shipping and the transportation of supplies, are enterprise-focused. For instance, in a manufacturing-based supply chain, inbound and outbound logistics are centered on the firm as a production function. Without exception, production-based SCM and the associated enterprise-focused logistics will be a fundamental part of this book. But in addition, as a unique feature and added value contributed to SCM by this book, the authors study trade-based SCM and the associated port-focused logistics, as will be further described below.

Verifying the fact that globalization is no longer a prediction, global supply chains are evolving from production-based to trade-based systems, and the associated logistics from the aforementioned enterprise focus to a port/airport/dry port focus (simply referred to as port-based). The relationship between enterprise-focused logistics and production-based SCM is as striking as it is intriguing. On the one hand, with regard to the supply chain as a business platform for the acquisition and provision of supplies and goods, enterprise-focused industrial *logistics* is inseparable from SCM. On the other hand, SCM contains production operations that are independent of logistics, such as manufacturing operations. In this sense, logistics can be clearly distinguished from SCM.

However, the production-based framework of SCM falls short in explaining recent developments in the real world, especially the so-called 'Walmart Model,' in which a 'factory' is a virtual global logistics network of multiple international manufacturing and service firms. This complex network of global logistics is a prototype of the trade-based global supply chain. The success of the Walmart Model rests on dynamic innovations in two key dimensions, namely global sourcing and contract manufacturing, and integrated multi-mode transport logistics. For instance, managing transport utilities and facilities, such as ports and airports, has become an inevitable part of trade-based SCM and port-focused logistics, especially in an international orientation. Supply chain logistics is precisely based on the intriguing interrelationships between production-based and trade-based supply chains, and between enterprise-focused and port-focused logistics. Following the convention adopted by industrial practitioners as well as scholars, logistics is assumed in this book to be enterprise-focused and SCM to be production-based, unless otherwise specified. In this regard, the definition of logistics by the Council of Logistics Management in 1998[1] helps explain the relationship between logistics and SCM:

'Logistics management is that part of supply chain management that plans, implements, and controls the efficient, effective forward and reverses flow and storage of goods, services and related information between the point of origin and the point of consumption in order to meet customers' requirements.'.

- The 'point-of-origin' herein is understood as a reference point after the inbound stage. When logistics is viewed as a subset of SCM, the production-based functions of SCM that are not considered in logistics are:Inbound flow management: strategic supplier alliance, outsourcing;
- Integrated supply-chain operations: adaptation planning and competitive coordination;
- Supply chain design concurrent with product development;
- Customer relationship management.

3.1.3   Innovations in firm-focal supply chain management: just-in-time production

There have been two basic supply management principles: push (i.e. make-to-stock) vs. pull (i.e. make-to-order).

Make-to-stock (MTS) production calls for a stock of standardized products to be produced before customer orders of the products occur (or arrive). A typical US firm using material requirements planning (MRP) system would operate under the MTS scheme.

Make-to-order (MTO) production, on the other hand, is only in response to customer (or upstream) orders in a just-in-time manner. The well known *kanban* system of Toyota production (to be further discussed below) is a classic example of MTO production, where any works at any workstation must be triggered (pulled) by a *kanban* signal that signifies the arrival of a new upstream order.

The key reason to have two different production systems is because manufacturing capacity is often limited, and that it always takes some time to complete an order (unless an infinite production capacity can be maintained which is, in reality, difficult if not impossible to achieve). Thus, in order to meet demands in a timely fashion, a manufacturer must either produce in advance (MTS) or produce in time (MTO). Although there is yet a scientific proof of this, industrial practitioners generally characterize MTS as suitable for 'selling old products to new customers,' while MTO is more suitable for 'selling new products to old customers.' Due to revolutionary advances in IT and constant innovation in manufacturing, there has recently been an increasing trend toward MTO production, such as production at Dell Computers. Supply capacity is measured in terms of production rate (i.e. volume produced per unit time) or unit production time (i.e. the reciprocal of production rate), since production always involves labor, and production capacity bears a behavioral nature, such as learning.

According to Ohno (1978), Toyota started the so-called 'Toyota Production System' (TPS) in 1945 when Kiichiro Toyoda, its then-president, determined that Toyota must 'catch up with America in three years. Otherwise, the automobile industry of Japan will not survive.' Although Toyota did not succeed in meeting this immediate objective, its president's ambitious campaign sparked the most influential innovation in the world of manufacturing, i.e. the birth of TPS, which rests on two pillars, as described by Ohno (1978):

1. Just-in-time (JIT).
2. *Jidoka.*[2]

An early implementation of JIT at Toyota was the well known *kanban* production system. On the contrary, *jidoka* did not attract as much attention. This was, in part, due to the fact that the concept represented a unique culture, or tradition, developed at Toyota and not as recognizable as JIT. *Jidoka* originated from the design feature in the invention of an automated loom in 1902 at Sakichi Toyoda (the predecessor of the Toyota Group). With this design, the loom would immediately stop if a thread broke. In the 1970s, Taiichi Ohno, a production engineer at Toyota Motor Co., institutionalized this idea of stopping production at every defect as *jidoka*. An early implementation of *jidoka* at Toyota was the self-activated stopping policy: a production unit or team at any stage of a manufacturing process was authorized to interrupt and stop the entire production line if a problem or failure occurred at that stage. Later, the concept referred to any mechanism of pre-authorized internal/local ruling and control. One should note, however, that it has since gone far beyond the self-stopping policy, just as JIT has gone far beyond *kanban*.

If one compares TPS with the typical industrial system in the US, two manufacturing principles can be identified, namely JIT and just-in-case (JIC) from Japan and the US, respectively. As mentioned earlier, JIT, stemmed from *kanban*, aimed at increasing speed, while JIC, based on the MRP system, focuses on safety. In JIT, the key to success is to eliminate waste in production. Therefore, it stresses waste reduction in inventory and set-up and promotes direct and strong links with all parties along the supply chain. The crusade against waste also entails a higher order of speed and accuracy, engaging *quality* (and quality management) as a unified and ultimate measure of performance. On the contrary, to succeed safely, JIC focuses on integrated (or centralized) planning ahead of time, relying on reliable forecasts and precautionary measures like safety stocks. Just as mass production is associated with the US automobile industry, 'lean production,' coined by Womack et al. (1990), characterizes the JIT system. Table 3.1 compares the characteristics of JIT and JIC.

As indicated in Table 3.1, the two systems underpin different philosophies and thus operational strategies: a pull strategy under JIT, and a push strategy under JIC. In short, production in a pull system is triggered (i.e. pulled) by demand (e.g. customer orders), while production in a push system is planned in advance (e.g. planned order release in MRP). These two strategies result in very different manufacturing approaches: make-to-order for JIT, and make-to-stock for

Table 3.1   The characteristics of just-in-time (JIT) and just-in-case (JIC)

Category	JIT	JIC
Philosophy	Quick win, reactive	Safe win, proactive
Strategy	Pull: eliminating waste	Push: planning
Approach	Make-to-order	Make-to-stock
Planning and control	*Kanban*	MRP
Critical measures	Reduction of inventory, set-up, lead-time, delivery time	Accuracy of forecasts, safety stock, lot sizing, service level
Competitive edge	Robust, IT simplistic, close-looped	Integrated, IT advanced, open-looped

Source: Authors.

JIC. Although their philosophies and strategies contrast drastically, both allow the use of common countermeasures, as mentioned in inventory literature. Nowadays, one can easily identify a JIT system that contains safety stocks, while it is equally common to find a JIC system that adopts elements originating from *kanban*.

3.1.4 Innovations in firm-focal supply chain management: the supply chain ring

Intuitively, supply chain, by its name, is an open system starting from the supply end, moving through transformation and ending at distribution and delivery. Recent advances in SCM are trended towards continued management after sales, notably customer relation management (CRM) and reverse logistics. To achieve this, i.e. to effectively manage the complete supply chain system, requires a great deal of adaptive integration. Hence, supply chain quality improvement is growing into an effective method for adaptive integration, and with wideband data (WBD) integration of the complete supply chain, a 'supply chain ring' emerges (Figure 3.3).

3.1.5 Firm-focal supply chain logistics

For the sake of the proposition, the authors cover the key logistical components of firm-focal supply chain management through a prototype example of the newsboy problem, as illustrated below:

A firm buys Q units per year of a product from an existing supplier, and sells to the market with an annual demand D. The firm buys at a cost of c per unit, and sells at a price of p per unit. That is, the revenue from selling one unit is p − c. Each unsold product at the end of a year will be salvaged for a discounted value of $c_0 \leq c$. **Question: *Determine quantity Q at the beginning of each year so that average annual profit can be maximized.***

A supply order Q is in fact a forward contract that will be either fulfilled (fully or partially) or defaulted on within a finite time period. The order is forward in the sense that it is designated for future needs and uses. The total time elapsed from the release of an order to the delivery of the order is termed the total supply lead-time, or simply the *lead-time*, denoted as L. The lead-time includes the time to acquire necessary supplies, the time to fill the order and the time to transport and deliver the order. Thus, the lead-time is the time lag between the release and delivery of an order.

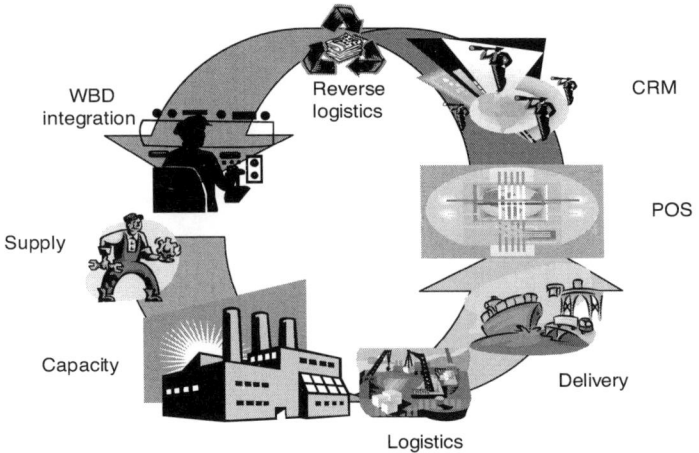

Figure 3.3 The supply chain ring: a circle of selling and buying
Source: Authors.

A manufacturing firm system differs from a service firm system in how production is triggered by the arrival of demands. For instance, television sets can be produced before the actual buyers are known, while (in most cases) medical practitioners cannot treat patients before their arrival at the clinic. In MRP, production orders are released in advance of the demand order arrivals, while *kanban* orders can be released only upon or after demand order arrivals. However, for a service order (like a clinic visit) the service operations can be performed only upon or after a customer's arrival. However, the preparation of raw materials for production in both manufacturing and service systems can (and should) start before the demands; the two differ in the implementation of the production process.

A supply order Q can be specified by (or converted to) the required workload. A supply order in general is expressed in terms of *base order* and *safety stock*, in a linear form of $Q = q+ss$ where q represents a base order and ss denotes the safety stock, which is allowed to be negative. The base order is determined by expectation (e.g. average demand over an average lead-time), while the safety stock is used as a countermeasure against uncertainties in demand. In manufacturing, safety stock is based on the consideration of the *service level*. A common definition of the service level is the probability of demands being met on time. Intuitively, a higher service level incurs a higher degree of satisfying customer needs, which requires a higher level of safety stocks. Also, it is recognized that a higher degree of variability in demand will require a higher level of safety stocks so as to achieve a given service level. In this regard, JIT promotes eliminating or minimizing the use of safety stocks with the condition that the service level should not be undermined.

Production quantity, actual sales and inventory status: While supply quantity Q must be determined before the realization of next year's demands, actual sales will depend on the actual demand D next year. Here it is important to note that actual sales may be different from actual demand. Specifically, the amount of actual sales would be the smaller of supply quantity and demand volume. That is:

Actual sales $= \min\{Q, D\}$.

When the supply ordered is greater than the actual demand ($Q \geq D$), there will be an overstock, expressed as:

$(Q - D)^+ = \max\{0, Q - D\}$

However, when $Q \geq D$, it is called out of stock (or simply stockout), expressed as:

Stockout $= (D - Q)^+ = \max\{0, D - Q\}$.

Probabilistic profits and costs: For given Q, the expected profit can be computed as:

$$E(\text{profit} \mid Q) \; E(\text{profit} \mid Q) = E(\text{sales profit} \mid Q) - E(\text{overstock cost} \mid Q) - E(\text{stockout cost} \mid Q)).$$

Now let us take a short break and refresh our knowledge on basic probability theory. The demand in the previous example is referred as a discrete random variable, as it only takes on a set of discrete values, such as $X = \{\cdots, x_1, x_2, \cdots\} = \{x_k\}_{k=-\infty}^{\infty}$. The probability measure of each possible demand realization is given by a *probability mass function*, as follows:

$f(x_k) = \Pr(D = x_k)$,   for all $x_k \in X$.

Then a cumulative distribution function can be defined as:

$$F(x_k) = \Pr(D \le x_k) = \sum_{i=-\infty}^{k} f(x_i), \quad \text{with } F(x_\infty) = \sum_{i=-\infty}^{\infty} f(x_i) = 1$$

However, one should note that the definitions, as introduced above, could merely be considered for the notational convenience. The expected profit under a given order quantity Q can be generally expressed as:

$$E\ (net\ profit\,|\,Q) = E(sales\ profit\,|\,Q) - E(overstock\ cost\,|\,Q)$$

$$= r \sum_{k=-\infty}^{Q-1} k \cdot \Pr(D = k) + r \sum_{k=Q}^{\infty} Q \cdot \Pr(D = k) - c_0 \sum_{k=-\infty}^{Q-1} (Q - k)\Pr(D = k)$$

$$= (r + c_0) \sum_{k=-\infty}^{Q-1} k \cdot f(k) + rQ\,(1 - F(Q-1)) - c_0 QF(Q-1)$$

The profit-maximizing solution of the prototype newsboy problem, Q^*, can be derived from the principle of *marginal equilibrium*: the expected marginal sales profit and savings should be equal to the expected marginal overstock inventory cost. That is:

$$r \cdot \Pr(\text{marginal sales profit}\,|\,Q^*) = c_0 \cdot \Pr(\text{overstock cost}\,|\,Q^*).$$

The 'marginal revenue' is the additional revenue generated by producing one extra unit (i.e. a total of $Q^* + 1$ yearbooks produced). It can be verified that:

$$\Pr(\text{marignal revenue}\,|\,Q^*) = \Pr(\text{selling one extra yearbook}) = \Pr(D > Q^*)$$
$$= \Pr(\text{selling one extra yearbook given } Q^* + 1 \text{ yearbooks produced})$$
$$= \Pr(D \ge Q^* + 1) = 1 - F(Q^*).$$

Similarly, it is possible to verify overstock that:

$$\Pr(\text{overstock}\,|\,Q^*) = \Pr(\text{one extra overstocked given } Q^* \text{ yearbook produced})$$
$$= \Pr(D < Q^* + 1) = \Pr(D \le Q^*) = F(Q^*).$$

Thus, the marginal equilibrium can be equivalently expressed as:

$$r\left(1 - F(Q^*)\right) = c_0 F(Q^*)$$

Then one can obtain the following condition for Q^*:

$$\Pr(D \le Q^*) = F(Q^*) = \frac{r}{r + c_0}.$$

In reality, demand can be, and often will be, met continuously, i.e. the set X, on which a random variable is valued, consists of continuous intervals. In this case, the *continuous* random variable is introduced, of which it takes value on a continuous interval on the real axis.

Service level and safety stock under continuous demand: A newsboy is faced with an uncertain demand D for the next period (e.g. day, week). The demand D can take any value over a real axis $(-\infty, \infty)$, with a certain probability distribution $F(x)$, a mean μ, and a variance σ^2, which are defined respectively as:

$$F(x) = \Pr(D \le x), \ \mu = E(x), \quad \text{and} \quad \text{var}(x) = E(x - E(x))^2$$

A continuous representation of the annual demand will lead to a normal random variable D, denoted as:

$$D \sim N(\mu, \sigma^2).$$

In inventory theory, the distribution function $F(x)$ coincides with the concept of *service level*, which is defined as the probability of *in-stock* inventory. That is:

$$\Pr(\text{adequate inventory to meet the demand}) = \Pr(D \le S),$$

where S is the on-hand inventory available at the time of occurrence of the demand D. Thus, under a planned inventory level (a decision variable) S, the resulting service level is measured by the distribution function with $x = S$. That is:

$$\Pr(D \le S) = F(S).$$

The newsboy needs to decide on the quantity of newspapers to purchase from the publisher for the next period. In addition to the unit revenue r and the overstock cost c_o, he should allow an intangible understock cost of c_u for each lost sale due to insufficient stock. If the demand for the next period is known for certain, then he can avoid both costs c_o and c_u. However, since the demand is now uncertain, the actual over- and understock costs will depend on the order quantity Q. A larger Q tends to reduce the possible understock costs while simultaneously increasing the probability of being overstocked. It is thus problematic to avoid over- and understock at the same time. In this case, how many should he order for the next period so as to maximize the expected net profits? For the continuous demands, a profit-maximizing newsboy orders quantity Q^* can be obtained from the optimal service level equation. That is:

$$F(Q^*) = \frac{r + c_u}{r + c_0 + c_u}.$$

When $c_u = 0$, the above equality becomes identical to the one that was obtained previously for the case of discrete demands. If there is no variability in demand, then it is logical to set the base order equal to the expected demand with no safety stock, which would mean a 100 percent service level. However, in reality, the world always changes, and thus the challenge lies in how to respond to uncertainty and variability. Demand variability tends to lower the service level, i.e. the probability of meeting the demand, and thus safety stock is amended JIC so as to maintain a desirable service level. Clearly, the consideration of safety stock level is related to the variability of underlying processes and the desirable service level. The safety stock in this case can thus be quantified, as follows:

$$ss = \text{safety stock} = z \cdot \sigma_e = \text{the amount in access to the mean}$$

where z is a safety factor that can be determined from any given service level, and σ_e is a variability measure, usually the standard deviation of forecast error (or the standard deviation of the demand σ if forecast error is not viable). If demand is a constant, there would be no variability ($\sigma_e = 0$), and the safety stock would vanish (i.e. $ss = z \cdot \sigma_e = 0$).

In summary, safety stocks are directly proportional to the variability of demand such that for a given service level, higher safety stocks are needed when the variability is higher. On the other hand, given the same variability, higher safety stocks are needed if a higher service level is required. If service level is taken as a common measure of manufacturing process quality, the ultimate challenge in manufacturing nowadays is to maintain a desirable quality level while using as little safety stock as possible. In fact, JIT centers on the very challenge of achieving the highest possible service quality level with minimal, or ideally zero, safety stock.

3.1.5 Supply chain governance

The dynamics of SCM are formulated by two categories of modeling framework: an inventory-production system, especially in the newsvendor context (e.g. Cachon, 2003) and a queuing type of service system (e.g. Kim et al., 2010). Hence, the efficiency measures of supply chain governance have been considered as the coordination under a price-quantity (or quantity-quantity) contract with certain incentive schemes, including inventory buybacks (Pasternack, 1985; Emmons and Gilbert, 1998; Lee, 2001), supply quantity discounts (Cachon, 2003), revenue sharing (Cachon, 2003; Cachon and Lariviere, 2005), two-part tariffs (Lariviere, 1999), quantity flexibility contracts (Tsay, 1999) and sales rebates (Taylor, 2002). In this sense, the measures on governance (e.g. the cost of governance) are largely omitted from the firm-focal supply chain models.

This approach, of production-based supply chain coordination, remained the mainstream in SCM research until Williamson (2008), who advocated the study of transaction-based supply chain coordination and governance through the lens of contract, under the theory of the firm as a governance structure (as opposed to the firm as a production function). By the TCE of SCM, coordination (and thus efficient governance) of supply chains can be attained in terms of both production cost minimization through ex-post efficiency measures, such as production input selection, and transaction cost minimization through ex-ante efficiency measures, such as organizational integration (Williamson, 2008). The key attributes of a transaction are therein identified as asset specificity, uncertainty and frequency; while the key attributes that define a

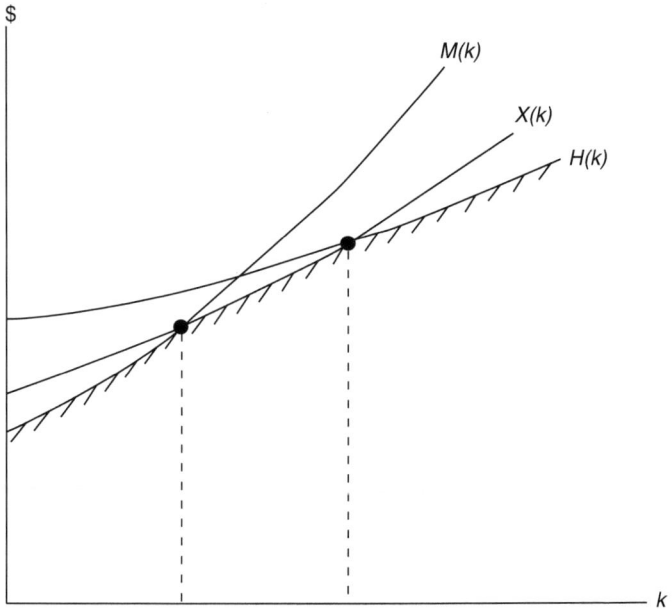

Figure 3.4   A heuristic model: the costs of governance
Source: Williamson (2002).

governance structure include incentive intensity, administrative control and the contract law regime, which are, in turn, interrelated with the attributes of alternative modes of governance, such as the competitive (market) mode suited to autonomous adaptations at one end, and the integrated (hierarchical) mode suited to coordinated (or systemic) adaptations at other end. As illustrated in Figure 3.4, *k* denotes asset specificity, while *M*, *H* and *X* denote three modes of governance, namely *market*, *hierarchy* and *hybrid*, respectively.

3.2   Outsourcing and global supply chains

Outsourcing refers to a firm that contracts out some, or all, of its production activities to domestic or foreign suppliers, as opposed to keeping them in house. In this section, the authors confine the discussion to global outsourcing, stylized in the North–South two-country trade framework under the trade theories, including the new trade theory (NTT) (e.g. Ottaviano, 2011). The containerization of shipping since the 1960s (see Chapter 2) has enabled the globalization of outsourcing and contract manufacturing, where a firm in the 'industrialized North,' which has developed a new product, is to contract the production to a supplier in the 'developing South' that is capable of acquiring the necessary manufacturing technologies, as exemplified by China becoming 'the world's factory' (see Ng and Tam, 2012). Such revolutionary movements have led to the current predominance of container ports and port/airport-enabled multimodal transportation, collectively referred to by the United Nations Conference on Trade and Development (UNCTAD) (1994) as the third generation of ports as an industry, where ports are broadly defined to include seaports, airports and dry ports/inland terminals. Nowadays, modern international business and trade are formed, in accordance with the Walmart Model, in terms of global outsourcing and contract manufacturing, with containerization as the enabling technology and ports as the necessary enabling infrastructure. For instance, the Walmart Model operates on a global network of contract manufacturers, together with a direct distribution network of supermarkets of its own brand. On the other hand, a firm-centric supply chain is homogeneous in the sense that it consists of firms in the same industry, each firm executing a production function.

3.2.1 Outsourcing: transaction cost economics

In essence of industrial ordering and supply provision, the mainstream SCM works are, so far, underpinned by the 'science of choice' approach (e.g. the theory of the firm as a production function; see Williamson, 2002), as opposed to the 'science of contract' approach. The latter approach is masterfully crafted and advanced by Williamson in two of his much celebrated heuristic masterpieces, entitled *The Theory of the Firm as Governance Structure* (Williamson, 2002) and *Outsourcing* (Williamson, 2008). In the former work, Williamson argued that:

> Economics throughout the twentieth century has been developed predominantly as a science of choice. Choice has been developed in two parallel constructions: the theory of consumer behavior, in which consumers maximize utility, and the theory of the

firm as production function, in which firms maximize profit. (Williamson, 2002)

The science of choice treats a firm as a bilateral monopoly (e.g. supplier-manufacturer) system that generates outputs with viable inputs under certain laws of technology, and focuses on the allocated efficiency of endogenous choice, mainly on how changes in prices and available resources influence quantities. In this connection, the alternatives of choice are quantified with regular input factors under the theory of the firm as a production function. However, this approach as a lens through which economic phenomena are studied is often preoccupied with optimization tools, and is therefore not always the most instructive lens (e.g. Buchanan, 1975; Williamson, 2002, 2008). The consequence is that SCM research has come through a surprisingly similar analytical path, evolving as an intrinsic domain of optimization-based disciplines of operations management and management science. On the other hand, the science of contract, coined in the mid-twentieth century by Buchanan (1964, 1975), no longer describes a firm as a stand-alone black box but as a 'sketched diagram,' i.e. an organization with alternative modes of governance subject to the underlying laws of technology. In contrast with mechanism design and the agent theory of the firm, it associates a firm with three attributes which are irregular (exogenous) factors, namely *incentive intensity, administrative control* and *contract law regime*, and are excluded from the models of the firm as a production function. However, this approach was largely neglected until Williamson's construction of the lens of contract streamlined under TCE. With reference to a specific class of transaction, he examined the contract/governance approach in the context of a make-or-buy decision under private ordering:

> Should a firm make an input itself, perhaps by acquiring a firm that makes the input, or should it purchase the input from another firm? [Accordingly,] the ultimate unit of activity...must contain in itself the three principles of conflict, mutuality, and order. This unit is a transaction. (Commons, 1932, p. 4; Williamson, 2002, p. 275)

Asset specificity (which gives rise to bilateral dependency), uncertainty (which poses adaptation needs) and frequency are the key attributes of transactions, which incur different transaction cost consequences (highly nonlinear) under different modes and attributes (discrete) of the governance structure. For example:

The requisite mix of autonomous adaptations and coordinated adaptations varies among transactions. Specifically, the need for coordinated adaptations builds up as asset specificity deepens. (Williamson, 2002, p. 275)

For easy reference, one should recall Williamson's heuristic model (Figure 3.4).

3.2.2   Transportation-facilitated outsourcing

The authors develop a transaction-engaged supply chain frontier (TESCF) model, which incorporates supply chain transaction attributes into the classical production frontier. Also, the authors construct a transaction function with inter-firm and inter-stage transaction attributes as input factors, in which transaction attributes are grouped into two categories, one of technical transaction factors of asset specificity and the other of non-technical transaction factors of *environment homogeneity* (e.g. geographical, demographic and institutional features). The transaction function is a modulation function with measures of transaction attribute as 'irregular' input factors, which are irregular in the sense that they are inter-stage, inter-firm and possibly intangible. In addition, there introduced is another category of inter-firm governance structure indicator (e.g. governance mode). In the TESCF model, the inter-firm governance indicators can be regulated by a social/non-commercial planner (e.g. public port authority) which is an inter-stage neutral party.

Transaction-embedded outsourcing system: Defining a supply chain frontier to be a composite frontier that maximizes supply chain output with minimized supply chain costs, including both production input costs and inter-stage transaction costs, the authors construct a TESCF model for a transaction-embedded retailer-supplier supply chain (Figure 3.5).

The retailer–supplier contract process can thus be streamlined (Figure 3.6).

To construct a framework of supply chains, the production function is assumed to be of the Cobb-Douglas type (e.g. Yan et al., 2009) with a regular input bundle denoted by a scalar input variable x with a rate of return scale $\alpha \geq 0$, that is, $g(x) = ax^{\alpha}$. Let $w \cdot x$ represent the regular production cost, with $w \in R_{+}^{m}$ as the unit regular input cost. Consider a regular transaction cost function, denoted by $\varphi(w, x, \kappa)$ (e.g. the costs of transaction inventory, warehousing and transportation), and an irregular transaction cost function,

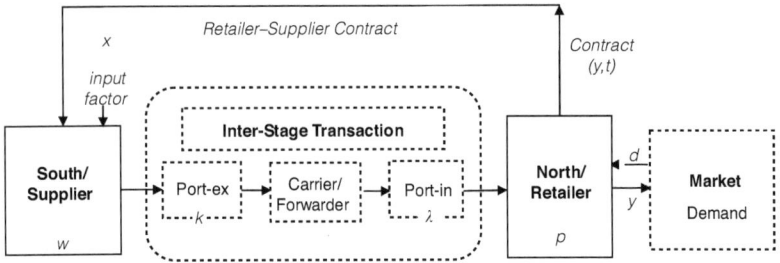

d: market demand;

y: contract quantity (task) by retailer, with an exogenous retail price p;

x: production input amount needed by supplier, with an exogenous input price w;

κ = indicator of governance mode at supplier (e.g., the number of operators in the port to indicate the mode of port governance. That is, $\kappa = 1$ for Singapore to indicate a centralized governance structure, $\kappa = 0.5$ for Shanghai to indicate a mixed mode of state-owned and private operators; while $\kappa \geq 2$ for Hong Kong to indicate a highly decentralized governance structure;

λ: indicator of governance mode at retailer.

Figure 3.5   Transaction-facilitated supply chain logistics
Source: Authors.

denoted by $\phi(y, z, \kappa)$ (e.g. governance and infrastructural costs). It begins with a basic transaction cost frontier (TCF) model without including regular transaction costs $\varphi(w, x, \kappa)$, as follows. For any production level $y \in R_+$, transaction attributes $z \in R_+^2$, and governance mode $\kappa \in R_+$,

$$
\begin{cases}
J(y, w, z, k) = \min_x \quad w \cdot x^t + \phi(y, z, k) = \sum_{j=1}^{m} w_j x_j + \phi(y, z, k) \\
s.t. \qquad \begin{aligned}[t] x &\in L(y; z, k) \equiv \{ x \mid f(x; z, k) \\ &= A(z, k) \cdot g(x) \geq y; \text{ for } y \geq 0 \} \end{aligned} \\
\qquad x \in R_+^m, \quad w \in R_+^m, \quad y \in R_+, \\
\qquad z \in R_+^2, \quad k \in R_+
\end{cases}
\tag{3.1}
$$

where $z = (z_1, z_2)$, and z_1 denotes the degree of asset specificity and z_2 stands for the degree of environment heterogeneity.

At infrastructure design stage: λ,κ
➤ Retailer: Governance structure for retail product generation
➤ Supplier: Governance structure for contract manufacturing
➤ Port (inter-stage): Supply chain inter-stage governance structure

At contract design stage: $J(\lambda), L(\kappa)$
➤ Retailer: Contract mechanism design; new product launch (pricing)
➤ Supplier: Production technology improvement; cost minimization technology
➤ Port (inter-stage): Port-focal supply chain contract design

At contract execution stage: $u(y;\lambda), x(y;\kappa)$
➤ Retailer's decision $u(y;\lambda)$: Determine contract quantity (task) and duration to maximize revenue over the contract period
➤ Supplier's decision $x(y;\kappa)$: Determine input quantity to produce to the contract with minimized production cost
➤ Port's function: Facilitate transaction-cost minimizing logistics (traders, shippers, carriers, forwarders, and 3PL)

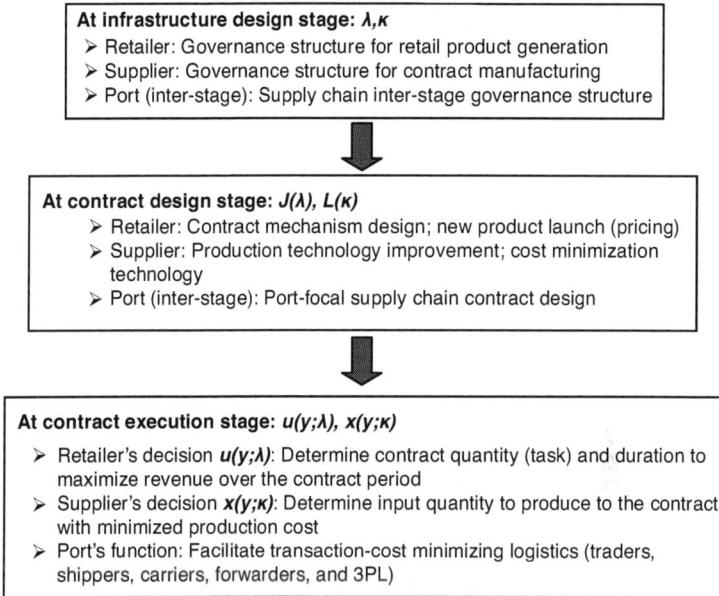

Figure 3.6 The retailer–supplier contract process
Source: Authors.

3.2.3 Global outsourcing strategies

There are two major operational strategies for global outsourcing, namely, foreign direct investment (FDI) and sub-contracting (also termed the 'arm's-length trade'), both strategies having ports as transaction facilities. For instance, the Intel Corporation adopts an FDI strategy: it assembles most of its microchips in wholly owned subsidiaries in China, Costa Rica, Malaysia and the Philippines. On the other hand, Apple subcontracts most of its iPhone production to independent producers in China. Both the outsourcing strategies require ports as transaction facilities between outsourcing resources and the global markets.

3.3 Port-focal supply chain and trade logistics

Port-focal production differs from firm-focal production a number of key characteristics. The differences can be summarized as follows:

- Port production technology is typically non-manufacturing, while firm production technology is mostly suited to manufacturing.
- In terms of organization structure, port production is engaged in a single-seller (port), multi-buyer (carriers) service system, while firm production can be characterized as a single-buyer (firm), multi-supplier manufacturing system.
- In terms of market and risk structure, port production is faced with idiosyncratic demands from the oligopolistic shipping market, while firm production is assumed to face systemic demands from a competitive market. In this regard, due to the high entry barriers to international shipping markets, the number of international carriers is small in comparison with the number of manufacturers around the world.
- Compared with firm-focal logistics, which is underpinned by an inbound-factory-outbound logistics chain, port-focal logistics integrates all modes of logistics and supply chains, including navigation, aviation and other types of transportation.

3.3.1   Transaction attributes of port-focal supply chains

In the context of the port-focal supply chain as the global mainstream, two categories of inter-firm transaction attribute are identified: (i) intra-firm *asset specificity*, which is firm-based (e.g. technology adoptability and interchangeability), as studied in transaction cost theories (Williamson, 2002); and (ii) *environment heterogeneity*, of an inter-firm and inter-stage nature (connectivity of supply chain stages, geographic and demographic characteristics, etc.), which is also a major focus of this book. To facilitate characterization of inter-stage and inter-firm transactions, the authors generalize the concept of port, from a production 'firm' as in mainstream SCM research to an inter-stage facility to engage supply chain transactions, including ports, airports, dry ports/inland terminals and logistical hubs. A port may consist of multiple operators (firms), including terminal operators and logistics operators, but by itself is a transaction facility engaging in certain transaction activities, such as inter-firm governance and infrastructure adoption, which are interactive with but exogenous to production inputs. For instance, more than 20 terminals of the port of Singapore are operated by one firm, namely PSA. On the other hand, the terminals of the port of Hong Kong are operated by several competitive firms (including DPW, HPH, MTL and PSA), each specializing in particular segments of the market, which is known as the horizontal competitive (specialized) mode of governance. The two container ports with the same production technology have adopted contrasting modes of governance: Singapore with a horizontal integration mode, and Hong Kong with a horizontal competitive mode. This contradicts both the theory of the firm as a production function (representing underlying production technology) and TCE with the firm as an intra-firm governance structure.[3] The ports of Singapore and Hong Kong both possess the same *asset specificity* measure in terms of their technical characteristics and the specifics of modern container port technologies. Through TCE, they should have adopted the same, or at least highly similar, modes of governance, either centralized integration or decentralized competition. Such strikingly different governance structures adopted at two of the world's top ranked container ports, with the same degree of asset specificity,[4] leads to the supposition that there are other causal factors missing from inter-firm characteristics.

3.3.2   Governance dynamics of port-focal supply chains and trade logistics

In line with the production function in classical efficiency theory, which characterizes production technology as a function of necessary 'regular' inputs (e.g. labor, capital and inventory), the authors construct a *transaction function* with inter-firm and inter-stage transaction attributes as intrinsic factors (uncontrollable) and with governance and infrastructure adoption as transactional input factors (as opposed to production input factors). The transaction function is a functional modulator to firm-based supply chain production which is represented by a production function with production input factors as its decision variables and input prices as system parameters. By the transaction nature as defined, transaction input factors are assumed not to be substitutable with production factors regardless of whether production factors are themselves substitutable or not. In the sense of incompatibility between transaction inputs and production inputs, transaction and production are considered to be interactive but not interchangeable.

Heuristic formulation of port-focal supply chain governance dynamics: With port-facilitated transaction and transportation, the authors put forth a heuristic formulation of port-focal supply chain governance dynamics, as follows:

1. Vertical (inter-stage across ports) transaction dynamics as a function of asset specificity: according to TCE, vertical specialization (competitive) mode of governance is transaction cost effective if asset specificity (z_1) is low; while a vertical integration mode is effective if asset specificity is high.
2. Horizontal (intra-stage) transaction dynamics as a function of environment heterogeneity: horizontal specialization (competitive) mode of governance is cost effective if the *environment heterogeneity* (z_2) is low; while horizontal integration mode is effective if the *environment heterogeneity* is high.

The port governance dynamics of port-focal supply chain and trade logistics are summarized in Table 3.2.

The heuristic formulation advanced in Table 3.2 is perfectly applicable to port governance, and reveals that inter-port *environment heterogeneity* is a key missing factor, as elaborated in what follows. Each port, as an inter-stage transaction facility, is inevitably associated with a dimension of *environment heterogeneity* which is exclusive from the existing supply chain models. Using the same example of port logistics, the above two ports, with the same degree of asset specificity, differ drastically in *environment heterogeneity*, Singapore

Table 3.2 The port governance dynamics of port-focal supply chain and trade logistics

Environment heterogeneity	Asset specificity low	Asset specificity medium	Asset specificity high
Low	V. Specialization H. Specialization Flat Convex TC	V. Mixed H. Specialization Steep Convex TC	V. Integration + H. Specialization Linear TC
Moderate	V. Specialization H. Mixed Steep Convex TC	V. Mixed H. Mixed Linear TC	V. Integration H. Mixed Steep Concave TC
High	V. Specialization H. Integration Linear TC	V. Mixed H. Integration Steep Concave TC	V. Integration H. Integration Flat Concave

Source: Authors.

being a free-port country neighboring several other countries with a high degree of geographical and demographic heterogeneity, while Hong Kong is a direct part of the hinterland logistics gateway of China, with a rather low degree of geographical and demographic heterogeneity. At the same time, the two ports (and their associated cities) also possess substantially different institutional and political systems, whereas the two hold the same degree of *asset specificity* as defined in TCE. Here one should note that *environment heterogeneity* is identified for the first time by the authors as an attribute of inter-stage transactions. How the two attributes affect the governance of logistics remains unanswered, which posts an active research question in transport, trade logistics and SCM. This issue, notably institutions as a factor of *environment heterogeneity*, its impact on the transformation of port governance, and the integration of ports into logistics and global supply chains, will be further discussed in Chapters 6, 7 and 8 with examples from both developed and developing economies.

Recent studies show that: (i) the efficiency of port-focal logistics is critically dependent on both intra-firm asset specificity (Williamson, 2002) and inter-firm *environment heterogeneity* (studied herein), with ports as transaction facilities. In particular, as evidenced in port logistics, horizontal integration (as compared with vertical integration) becomes more cost effective as *environment heterogeneity* increases, given the same degree of asset specificity within individual ports; and (ii) there is a logistics TCE that an adaptation advantage, in transaction productivity, for example, can be gained through the port-focal industrialization of supply chains, leading to geographically concentrated horizontal specialization and differentiation. In sum, the industrialization of global logistics brings about the ever-growing emergence of port-focal urbanization in logistics and supply chains. This raises the question: with such development, how exactly have ports evolved since containerization is invented? This question will be further investigated in the next chapter, on logistics, supply chains and port evolution.

4
Logistics, Supply Chain and Port Evolution

As mentioned in Chapter 2, containerization affected not only the development of shipping, but also other components of the transportation and logistical systems, including ports, and inland transportation. In this light, this chapter discusses how containerization, the restructuring of shipping and the development of global supply chains shaped the evolution and development of ports up until today. At the end of the chapter, we briefly discuss Hong Kong and the PRD ports in Southern China, which provide an example of the multi-directional development of ports towards port-focal logistics.

4.1 Ports as key components of maritime logistics and supply chains

Port was derived from the Latin vocabulary *portus*, meaning gateway. Traditionally, it was understood as facilities located along the sea-land interface to receive ships and transfer goods and passengers. Given this understanding, as illustrated in Figure 4.1, ports were always an isolated set of facilities and infrastructure, and nearly always maintained close connections with their forelands and hinterlands.

Indeed, being the interaction point between land and maritime spaces, ports traditionally served as the economic and cultural centers of cities and proximate regions. However, the recent technological advancement in shipping, increase in international trade and the global division of labor have fundamentally transformed

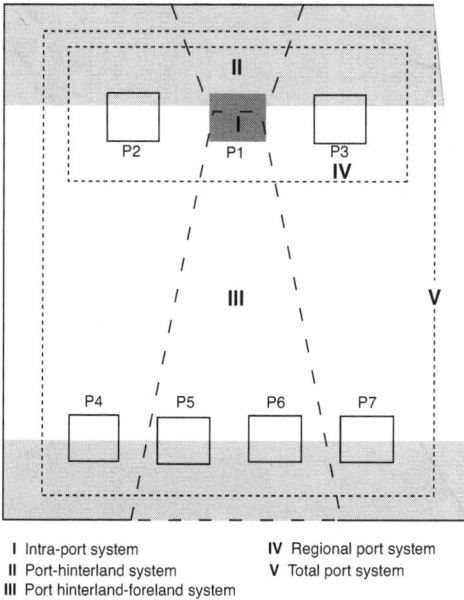

I   Intra-port system
II Port-hinterland system
III Port hinterland-foreland system

IV Regional port system
V   Total port system

Figure 4.1 The port system
Source: Authors.

the nature of ports. With increasing ship size, the restructuring of shipping networks and the development of mergers, acquisitions and strategic alliances, the number of ports started to reduce, trans-shipment traffic escalated and mega-sized containerships continued to call at fewer ports within each region. As mentioned in Chapter 2, this required shipping lines to operate more flexible schedules and this development was pivotal in changing the ports' role and landscape. Given its capital-intensive nature, requiring high load factors (≥ 80 percent) in order to benefit from economies of scale by using containerships with higher capacities (Slack, 1998), the service frequency and turnaround time of such ships had to be increased and decreased, respectively. This led to the establishment from multi-port-of-call to trunk-and-feeder shipping networks, especially along the trunk routes between East Asia, North America and Western Europe.

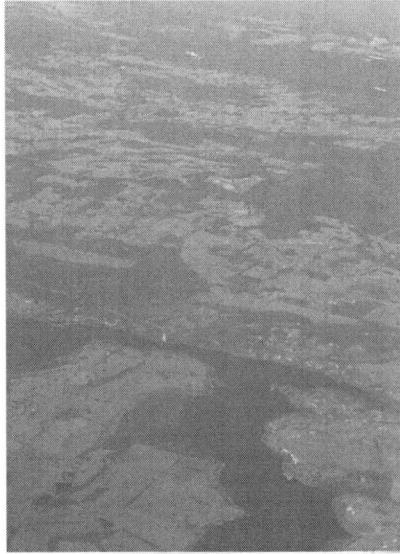

Illustration 4.1   Mega-ships sometimes face physical restrictions
Source: Authors, taken in Tasmania, Australia (2012).

However, while mega-ships theoretically enabled shipping lines to become better off in terms of unit cost through economies of scale, in practice, they needed to tackle additional, and substantial, challenges. In many cases, mega-ships proved to be more difficult to operate due to more demanding requests in terms of monetary costs,[1] time and physical restrictions, e.g. navigation channels along rivers and canals, berthing water draught, access channels (Illustration 4.1) and more demanding stevedoring facilities.

As a consequence, the optimal ship size should be decided not only by their operational costs at sea, but also by the negative externalities that their sizes could impose on other components of logistics and supply chains, especially ports (Jansson and Shneerson, 1982; Ng and Kee, 2008). In this case, ships with higher capacity were often constrained by ports' physical conditions (Figure 4.2), which could cause changes in the proportions of products received from various sources, thus posing problems to shippers and stopping certain ships from accessing locations where demand was present.

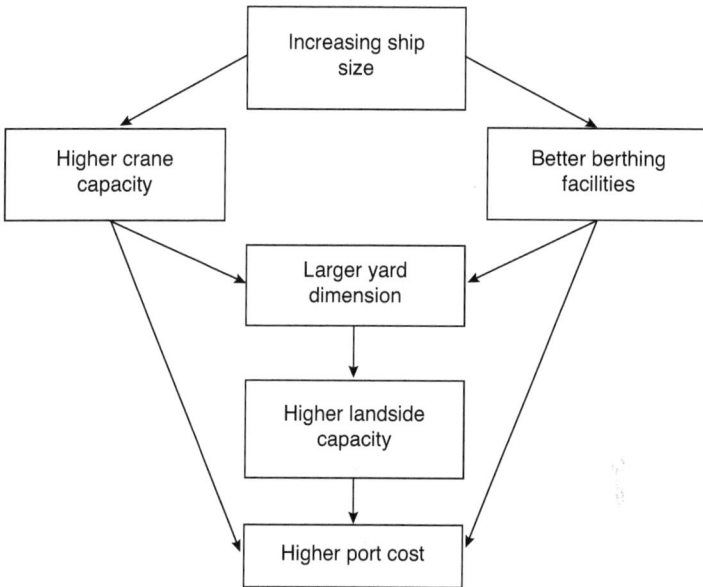

Figure 4.2   The relationship between increasing ship size and port cost
Source: Authors.

Increasing ship size would require higher port operational costs as the facilities required in handling such ships, e.g. crane capacity and berthing facilities, and thus ports' financial commitments, would increase at the same time. In this regard, integration would be pivotal, implying the extent to which different stakeholders could work together in an effective, collaborative manner to achieve mutually acceptable outcomes, commonly known as the optimal solution (Carbone and Martino, 2003), along the supply chains. While shipping lines invested heavily in ships (through either purchasing or chartering), the commitments by ports were equally substantial, if not more so, as efficient port operation was pivotal in reducing overall transport costs, thus improving the quality of the transportation and logistical processes. Mega-ships, with extraordinarily high capital costs, could not spend excessive time in calling at additional ports for more cargoes and incur the high port costs in serving them (Notteboom, 2002; Ng and Kee, 2008).

Substantial financial investments, both fixed and variable, in shipping ensured that cost minimization would become an important consideration during movement between production units and markets. Hence, it became extremely important for cargoes to be carried as smooth as possible, and with minimal detraction. In this situation, with ports acting as the sea–land interaction point, there was a pressure of 'integration' on ports to facilitate such development, and this brought forward the so-called 'port logistics' and 'maritime logistics hubs' (see Nam and Song, 2011). Under these concepts, apart from being a simple sea–land interface, ports should be integrated into supply chains on the basis of efficient physical cargo flows, common strategic goals and innovative organizational relationships. The physical flows consisted of the entry points, cargo loading and unloading, transit and storage and linkage systems, while information flow related to all relevant operational information concerning physical cargo flows, and each sub-system should be interconnected according to cargo flows within the system.

Ideally, ports within this system should provide not only general logistical services, e.g. stevedoring and warehousing, but also value-added services to meet specific demands, e.g. quality control, re-packaging, assembly, calibration, customization and re-export. The efficiency of the logistical entities largely depended on ports as it acted as the integrating and coordinating mechanism between different components, e.g. shipping lines, inland transportation and warehousing (Bichou and Gray, 2004; Miyashita, 2005). Since the general adoption of containers, ports had become vital components in ensuring the success of logistics, supply chain and JIT inventory management strategies, where increasingly demanding customers push service providers hard to provide speedy, just-in-time services at reasonable cost. In short, contemporary ports should play the brokering role of knitting different logistical sub-systems together.

As mentioned before, another implication of the recent technological and economic development of shipping was that it had considerably strengthened the bargaining power of port users, i.e. shipping lines. With the establishment of strategic alliances and the restructuring of major shipping networks, ships became more footloose in choosing ports of call and the traditional consensus that ports possessed certain natural hinterlands and thus few worries about losing customers became obsolete. Ports were gradually subdued by their users as increasingly borderless trade broke down discrete natural hinterlands and replaced them with 'common hinterlands' (Hayuth and Hilling, 1992; McCalla, 1999); this development had a substantial impact on port competition, and service quality became an increasingly important factor in deciding port competitiveness.[2] In the past two decades, shipping lines, especially those operating trans-continental services, have often exerted substantial pressure on ports to improve their intra- and superstructures so that they could economically benefit from deploying containerships with larger capacities (Notteboom and Winkelmans, 2001; Heaver, 2002).

Given such pressure, ports were forced to invest so as to improve service quality continuously but without any guarantee of getting new (or even retaining existing) customers. Thus, competition between major ports intensified, causing the traditional philosophy in port management, characterized by a strong bureaucratic structure, to become obsolete, especially since the traditional objective for ports to act as the bases to control hinterland markets during the colonial period in the early twentieth century (World Bank, 2007) had long gone. Not to be left out of the development of global supply chains, port operators needed to understand the attributes users considered when choosing a particular port. Many ports were forced to continuously improve their facilities and services in order to attract potential customers to use their services (for instance, see Meersman and Van de Voorde, 1998; Chang et al., 2008). Without doubt, ports specially designed for container handling had intensified the competitive environment for port services as containerization represented capital-intensive activities, and thus a substantial financial burden, to port authorities around the world (Heaver, 1993; Slack et al., 1996). To enhance competitiveness, port managers were forced to alter their focus from the traditional question of whether their ports could handle cargoes effectively to whether they possessed the ability to attract potential customers, fight off competitors and, in turn, be more responsive to the demands of users. Such requirements partly explained the increasing privatization of port operations in different parts of the world and several studies underlined this process (e.g. Juhel, 2001; Heaver, 2002).

4.2   Port management and governance reforms

As a strategically important component of global logistics and supply chains, ports had become subject to more complex demands. Port operators were confronted with various challenges according to the dynamic development of the global supply chain. Indeed, the task of integrating different components within the system while at the same time sustaining a high level of efficiency is not always easy for modern ports.

When certain exogenous circumstances started to change like those mentioned above, the operating environment also changed, thus causing substantial pressure, functional, political and social (Oliver, 1992), on the original institutional setting. The changing environment meant that ports, often characterized by bureaucracy in their pre-containerized stage, found it extremely difficult, if not impossible, to tackle the new challenges effectively, causing unsatisfactory outcomes, e.g. inefficiency and lack of competitiveness, and it is this external contingency that triggered most of the institutional reforms (Aldridge, 1999). They were faced with a choice – to drop out or to develop a new management structure and practice based on redefined objectives, and the evaluation of different reform tools. Throughout the past two decades, many ports, especially those that had ambitions to become strong(er) regional hubs, have undertaken significant reforms so as to be well equipped to tackle the rapidly changing economic environment, many of them gradually moving away from direct public management to increased private participation in port operations, thus creating more complex, hybrid organizational entities (Cullinane and Song, 2002; Brooks and Cullinane, 2007), i.e. the post-reform settings. Although reform objectives could differ due to various factors, notably the influence of existing institutional settings (Ng and Pallis, 2010), virtually all reforms shared common strategic goals – to enable the port to re-fit within the new environment – leading to satisfactory outcomes again, e.g. enhancing efficiency, deriving economic benefits through competition, minimizing bureaucracy, reducing public sector investment commitments, enhancing management skills and adopting more effective port labor organization. Encouraged by the move towards neoliberal economic objectives and policies, many countries and regions became much more open to foreign investment and participation in infra- and superstructures, especially after the end of the Cold War and the establishment of the World Trade Organization (WTO) in the 1990s.

Complementing this trend, there were substantial studies, both academic and commercial, investigating the approach, tools and models of port management and governance reforms (for instance, Heaver, 1995; Cass, 1998; Notteboom and Winkelmans, 2001; Ng and Pallis, 2010; Shou et al., 2011). While subject to necessary adjustments based on local circumstances and dedicated requirements, according to the World Bank (2007), post-reform settings can be summarized into four major models with reference to the extent of public sector's involvement in port operation and management, namely service, tool, landlord and fully privatized ports. During this process, the increase of private involvement was often the key objective, and this was usually linked to the (partial or complete) transfer of ownership of port/terminal assets from the public to the private sector, inviting the private sector to invest in port/terminal infra- and superstructures, or transforming an existing public agent into a private entity (like the transformation of port authorities into private (state) corporations, see Ng and Pallis, 2010). Complementing the philosophy of management reforms, generally speaking, the main objectives of increasing private involvement included the improvement of management capabilities and service quality, the reduction of public sector financial demands (thus sharing risks between public and private sectors), the stimulation of private (sometimes foreign) investments, and in some cases, wealth redistribution and the dilution of trade unions' power. Alternatively, ports could reform their practices and enhance service quality without significant changes in management structure or the transfer of port properties to the private sector. In such cases, any attempts to enhance service quality were usually made through modernization (such as better work practices within the bureaucratic constraints), commercialization of operational and management practices (such as accountability for own decisions and performance) or transforming the public authority into a quasi-private corporation. Indeed, the

Illustration 4.2   Rotterdam was one of the first ports in the world to corporatize its port authority

Source: Authors, taken in Rotterdam, Netherlands (2010).

corporatization of previously public port authorities could be found in some major ports around the world, e.g. Busan in South Korea, the Piraeus in Greece, Tianjin in China and Rotterdam in the Netherlands (Illustration 4.2), with commercial objectives, practices and development plans.[3]

In most cases, while operational aspects were transferred to private firms (be they terminal operating companies, shipping lines or other stakeholders), the public sector often retained its role as regulator and landowner, while in some cases it also retained the ownership of the infra- and superstructures of the ports (World Bank, 2007). In this regard, the establishment of 'autonomous ports' (*ports autonomes*) in France served as an example. Of course, there were also cases where the public sector transferred nearly all regulatory, land and operational functions to private firms, with the aim of withdrawing public involvement, and became fully privatized ports, as exemplified by the British port of Felixstowe. According to Ng (2009), the operator of the port of Felixstowe could even take up policing duties, if they chose to do so. However, given the perceived strategic importance of ports, both geopolitical and economic, in practice, fully privatized ports were seldom adopted (apart from Felixstowe, the British port of Tilbury was another notable instance).

Here a question could be raised as to how such models were established. In this regard, reform tools were necessary to make the reform process possible, notably the arrangements for port facilities to be transferred from the public to private sector smoothly, and the procedures and arrangements for private investments. The major tools for port management and governance reforms are illustrated in Figure 4.3.

A major tool for port management and governance reform was the outsourcing and/or managing contract. Under such an arrangement, an agreement with the lessee was made so as to confer on the latter the right to use particular asset(s), including infra- and/or superstructures. In return, they were liable to pay certain fees to the lessor. The lease usually lasted several years before any contract renewal discussions took place. However, in cases where private investments were encouraged, the tools employed were often more than just leasing infra- and superstructures. In this situation, concessions often came into force, Build-Own-Operate-Transfer (BOOT), Build-Operate-Transfer (BOT) and Build-Own-Operate (BOO) being the most popular options. Under BOT, private sector invested in terminal infra- and superstructures, as well as their operation, for an agreed period (usually more than 20 years). It was expected that the terminal, and its facilities, would be transferred back to the public sector when the concession period ended. In other words, concessions could be understood as the tool most often used for so-called 'partial privatization.' On the other hand, BOOT and BOO were similar to BOT except that, under the first arrangement, the private terminal operator also owned the terminal(s), and its facilities, during the concession period, while in the second arrangement, the operator did not need to return the terminal(s), and its facilities, back to the public sector afterwards. Indeed, concession

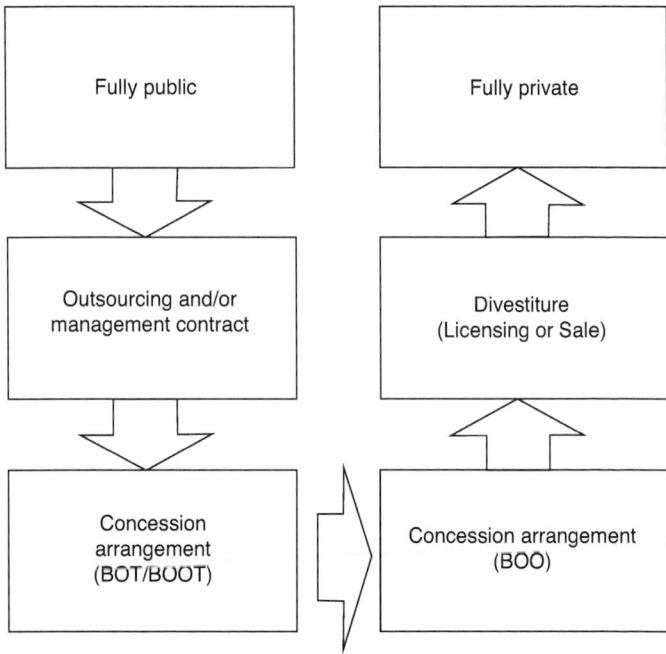

Figure 4.3 The major reform tools in port governance
Source: Authors.

has proved to be a very popular tool to encourage private investment and the operation of port terminals in both developed and
developing economies around the world in the past decades (Theys
et al., 2010).

However, under most concession agreements, the public sector
retained the ownership of the port's land, and the terminal operator(s)
usually had to pay rent. For example, the Port of Rotterdam Authority
(*Havenbedrijf Rotterdam*) (PoR), the management firm of the port of
Rotterdam, had to pay more than €45 billion (about US$50 billion)
worth of dividends annually to Rotterdam's municipal government as
rent, while PoR sub-leased the port's land to individual terminals (Ng
and Pallis, 2010). A comprehensive privatization involved the divestiture of public property, sometimes the port's land, to the private
firm, through either licensing or sales. Under such arrangements,

the private firm could even change its land use (depending on the divestiture conditions) for alternative functions, if they deemed it relevant. Thus, unsurprisingly, most governments around the world were not keen on the idea of full privatization. This issue will be further discussed in Chapter 6.

While the objective of port reform was to enhance service quality and thus competitiveness, this process led to the rise of (multinational) terminal operators from different parts of the world. This could be dated back to the mid-1980s when P&O Ports invested in Port Klang in Malaysia, soon followed by the Hong Kong-based firm Hutchison Whampoa Group (later known as HPH), and established its first European headquarters in Felixstowe. Other notable players included the Stevedoring Service of America (SSA), International Container Terminal Service (ICTSI) and Eurokai. This process accelerated in the 1990s, largely thanks to the success of the pioneers, the end of the Cold War, the massive wave of globalization and the increasing (arguably excessive) popularity of neoliberal ideology in economic policies (Harvey, 2005). Apart from HPH, a number of notable players entered the market, including DPW, PSA, CSX Intermodal Terminals (CSX), Hamburger Hafen und Logistik (HHLA), Dragados and Terminal de Contenidors de Barcelona (TCB). In Europe, for example, HPH, PSA and DPW had operations in various ports, including Antwerp, Felixstowe, Le Havre, Marseille (Fos), Rotterdam and Zeebrugge, to name but a few. Apart from companies dedicated to terminal operations, shipping lines participated in port terminal operations, with the presence of Maersk, China Ocean Shipping (Group) Co. (COSCO) and Mediterranean Shipping Co. (MSC) in various European ports, including Antwerp, Genoa, Le Havre, Marseille (Fos), Rotterdam, Tangiers and Zeebrugge. Indeed, the increasing prevalence of maritime logistics and supply chains since the 1990s, facilitated by neoliberal reforms which opened up many ports to foreign involvement and investment, prompted shipping lines to provide similar services, often as dedicated terminals serving their own ships (in some cases, also ships operated by other members of the strategic alliance that they belonged to). By gaining footholds on sea–land intersection points, many of them started offering multimodal transport and logistical services to their customers, i.e. shippers. For example, apart from port terminal operations in Bremerhaven, Rotterdam and Zeebrugge, Maersk operated the European Rail Shuttle (ERS),[4] providing freight rail services between the aforementioned ports and the inland regions of the European continent. On the other hand, until recently, partly complementing its operation at the Ceres Paragon Terminals in the port of Amsterdam, Nippon Yūsen Kabushiki Kaisha (NYK) was involved in offering various logistical and value-added services within the port, e.g. warehousing, inspection, sorting, labeling, repackaging and overland delivery by truck to other European destinations. Some shipping lines even provided one-stop logistical solutions to their customers, as witnessed by the establishment of logistical branches, Orient Overseas Container Line (OOCL) Logistics being an example.

4.3   The multi-directional development of ports

It was clear that while containerization had triggered port management and governance reforms, it had simultaneously opened up ports to foreign participation and investment. Hence, the reforms catalyzed the horizontal and vertical integration of terminal operators and shipping lines, and enabled shipping lines and other transport operators to control the supply chains. Such a transformation process was by no means a straightforward task and ports needed to undergo a self-reinvention process. This was certainly not helped by the fact that the changes in management and governance philosophy, and the pressure for them to be integrated into maritime logistics and supply chains, encouraged more international trade and transactions, thus causing congestion within port areas. In this regard, ports (and indeed maritime logistics hubs) should be transformed in an appropriate way so as to maintain their competitiveness and survival. This was especially in view with the intensified competition between nearby ports, the pressure for regional cooperation between proximate ports (Homosombat et al., forthcoming), as well as the dynamic transformation of production plants and regional development.

The first possibility was the infiltration of ports into inland regions and provinces, and the strengthening of the connection between ports and inland transport and logistics infrastructures. Indeed, the dynamics between containerization and port development were affecting each other, as would not be possible without the support of an efficient supply chain, with an unimpeded flow of cargoes, of which when translated in colloquial terms implying integrated transportation network, with the need for high quality management of cargo flows with low inventory costs and more reliable delivery and distribution. In this respect, supply chains, of which ports were important components, had to ensure that merchandise that they handled could sustain competitiveness within the global market, while the shipment process of cargoes must be smooth and economical so as to attract foreign investment (Sahay and Mohan, 2003). This was especially true for developing economies, such as India, Brazil and China. According to a study conducted by the World Economic Forum, rather than distance, the core attribute in determining the quality of global supply chains was connectivity, i.e. the smoothness with which freight, and thus trade, could be transported (World Economic Forum, 2008). Also, it noted that while distance accounted for 20 percent of variation in freight rates, competition and economies of scale had stronger impacts on transport costs, especially the trans-shipment costs triggered by the lack of direct connectivity.

In certain cases, high delivery costs were sustained by fragmented supply chains and poor logistical service levels and connectivity, thus trapping many developing economies into sustained poverty. In this case, various neoliberal management and governance reforms among major ports around the world, as mentioned earlier, offered an opportunity for ports to play bigger roles in strengthening their connections with inland transport and logistics facilities. As early as the late 1990s, Slack (1999) discussed the possibility of 'satellite terminals,' or dry ports/inland terminals, as a potential solution to the port congestion problem, and to enable ports to penetrate into inland regions so as to strengthen their connections with shippers and other supply chain stakeholders, or the so-called 'port regionalization' coined by Notteboom and Rodrigue (2005). In this regard, a dry port/inland terminal can be understood as an inland setting with cargo handling facilities to allow several functions to be carried out, e.g. cargo consolidation and distribution, temporary storage of containers, customs clearance and connection between different transport modes, allowing the agglomeration of both public and private institutions, which would facilitate interaction between different stakeholders along the supply chains. Dry ports/inland terminals conducted various functions similar to ports and provided several needs along the supply chains, namely aggregation and unitization of cargoes, in-transit storage, customs clearance, issuance of bills of lading in advance, relieving congestion in gateway seaports, assistance in inventory management and warehousing, and the deferment of duty payment for imports stored in bonded warehouse (Meersman et al., 2005).

Inevitably, however, this required synchronization and close cooperation between different stakeholders along the supply chains. Indeed, the direction of such a transformation process, notably the relationship between ports and dry ports/inland terminals, was never very clear, and thus what exactly maritime logistics hubs should be, as well as the necessary conditions for them to perform and remain competitive, remained rather ambiguous. Several important questions remained unanswered: how could one assess whether such an integrated supply chain would be efficient and competitive? Could its development independent from cargo stevedoring? What should be the roles of the different stakeholders and their interrelationships? Would substantial amounts of money be invested in this development justify market requirements? Little wonder, during the transition, that planners often found it difficult to identify new core business and develop appropriate re-positioning strategies, as they were largely unsure of the necessary steps to allow the aforementioned self-reinvention process to take off.

The second possibility was that ports under transformation should attract most of the logistical and value-added activities around the port area. The inclusion of logistical and value-added activities around the port area, e.g. free trade zones, warehousing and packaging, etc. would make ports more attractive to their users, while at the same time also enhance their rather weak bargaining power over users (as mentioned earlier) in the past two decades – the concept of port-centric logistics (see Chapter 1). However, while such a strategy possessed certain advantages, it also created some new problems. Notable advantages included the possible inadequacy (or even the lack of) coordination among different port stakeholders, of which it could paralyze the effectiveness of supply chains. As noted by Martin and Thomas (2001), development had altered the nature of the port and led to the rise of a new, but more complex, port community. Moreover, various management and governance reforms, and the evolved port communities, ensured that the port authorities did not necessarily dominate the port community and other port stakeholders, nor would always have a bigger say in port affairs, including development. Such a situation raised the question of whether reform could really achieve the objective of adapting to the changed environment, thus leading to a satisfactory outcome again. This led to more difficulties in managing the new structures, as noted in a number of studies (for instance, Notteboom and Winkelmans, 2001; Wang and Slack, 2000; Wang et al., 2004).

In this regard, as noted by Lam et al. (2013), effective stakeholder management would be the key to defining the success of such a development direction. An excellent example is Hong Kong and the PRD in Southern China. Being part of the trade-based economy, its port was historically regarded as one of the main pillars of Hong Kong's economic development. However, the global and regional maritime landscape underwent substantial changes. These included increasing trade between China and overseas markets (resulting from, *inter alia*, the 2005 abolition of quotas imposed by the US and Europe on Chinese textile exports), challenges from peripheral ports (notably Shenzhen and Guangzhou), the increasing volume of intra-Asian trade (like the '10+1' Asian Free Trade Zone), and the economic turmoil in 2008, which accelerated the industrial transformation of the PRD. The stated factors had challenged the expected forecasts regarding the current and future development of the port of Hong Kong, notably its shrinking hinterland size within the PRD (not helped by the intensified competition of its port terminals with nearby ports, notably Shenzhen). In the past decade, in terms of container throughputs (in TEUs), the port of Hong Kong's ranking had slipped from 1st to 3rd behind Singapore and Shanghai, while the annual container throughputs of Shenzhen had grown at least five times faster than Hong Kong. Hence, it needed to undergo strategic changes, notably its gradual integration within China's national and regional planning. Recently, Hong Kong, and its port, was included in the 12th National Five Year Plan (FYP), and the National Development and Reform Commission of China included the port of Hong Kong in *The Outline of the Plan for the Reform and Development of the PRD* (2008–2020) (Chapter 11, Article 2). Also, according to the Framework Agreement on Hong Kong/Guangdong

Co-operation (2010), signed between the HKSAR and Guangdong Provincial Governments, the port was expected to integrate within the PRD so as to help in establishing a system with the functions of different ports complementary to each other (Chapter 9, Article 4). However, until now, the direction of this transformation process is still rather blurred.

Of course, the situation was not helped by the changing industrial landscape of the PRD, which acted as the 'world's factory' for more than three decades since the late 1990s. Living costs rose rapidly in the PRD, forcing manufacturing firms to constantly increase workers' remuneration, while wages surged at an annual rate of 17 percent. Preferential policies offered by local authorities since the early Open Door Policy period have gradually been withdrawn from labor-intensive production, as municipal and provincial governments now aim to set aside more land and subsidies for high-tech and service-oriented industries. Hence, the advantages of the PRD for labor-intensive manufacturing faded rapidly, and this prompted substantial firms to find alternative locations in inland provinces – usually economically less developed but with abundant, cheaper labor and land – for their production plants. In this regard, the Chinese government implemented a strategic plan to promote economic re-balancing among the provinces in the hope that inland provinces could achieve sustainable growth in the long run, and to narrow income gaps between different regions (Qiao et al., 2008; Chen and Groenewold, 2011). Indeed, in China, both the national and the provincial governments offered generous incentives, like tax concessions, cheap land, and even free factory buildings, to encourage private firms to invest in inland provinces.

Such incentives further accelerated the relocation process and might pose significant challenges to the PRD ports, including Hong Kong, which had benefited substantially from the export and import growth driven by the manufacturing boom since the early 1980s. Their hinterland access costs would increase, leading to reduced traffic volume. In addition, if production plants were relocated to provinces close to other gateway ports, such as Shanghai, a substantial volume of traffic might also switch away. Since ports along the Yangtze River Delta (YRD) could offer competitive inland shipping services to several major inland cities and provinces, manufacturers relocated to these locations might find it cheaper to transport their cargoes to ports in YRD rather than to PRD, even if the distances to the ports were similar. Hence, the relocation of manufacturing operations based in the PRD would not only influence the performance of the PRD ports, notably Hong Kong and Shenzhen, but might also re-shape the competitive landscape between different port clusters in China. In this regard, Homosombat et al. (forthcoming) undertook an economic analysis on the situation and concluded that the port of Hong Kong must enhance its hinterland access, and thus its connection with the transport and logistics infrastructures located in China's inland provinces.

However, according to them, this measure would not be adequate to sustain the port of Hong Kong's competitiveness, and ensure its survival, in the long term. They further recommended that it should strengthen itself by providing a wide range of multimodal, logistical and value-adding services. Finally, they recommended that it should increasingly collaborate with nearby ports, notably Shenzhen, Guangzhou and other PRD ports, so as to form the 'Greater PRD Port Cluster,' which might effectively compete with other port and port regions, like the YRD ports. The discussion is still ongoing. Nevertheless, this implies that the development direction of the port should be multi-directional: port-centric logistics, the infiltration of the port into inland regions, as well as collaboration between proximate ports. Indeed, the situation of Hong Kong and PRD clearly highlights the importance of port-focal, rather than just port-centric, logistics, as mentioned in Chapter 3.

In sum, this chapter illustrated that recent developments in shipping forced ports to evolve. Ports were forced to undertake largely neoliberal management and governance reforms so as to enhance efficiency, the reduction of financial commitments from the public sector, as well as responding better to the demands of users. Such a transformation opened the door for ports to become more integrated into supply chains, while ports were also provided with an opportunity to transform themselves into maritime logistics hubs with a more hybrid community. However, until now, there is still no consensus on the best direction of development for such a maritime logistics hub: should ports encourage the concentration of logistical and value-adding activities around themselves, or should they strengthen their bonds with inland markets and transport and logistics infrastructures, especially dry ports, so as to sustain their attractiveness to users and, ultimately, competitiveness? Even within the reform process, given the realistic situation within different countries, notably diversity in political tradition, institutions and systems, would a similar solution, when applied to different cases, lead to the same outcome? As questioned by Ng and Pallis (2010), could a solution be generically applied to different ports around the world? With the potential pitfalls of increasingly governance complexity, was there a so-called 'best reform practice' that could fit the new circumstances and be globally applicable, or did the choice of appropriate reform approach actually depend on the regional and local circumstances? Of course, one of the best ways to start answering the above is to identify relevant methods so as to provide accurate and reliable benchmarks on the performance and efficiency of ports (and the port-integrated logistical system). These matters will be discussed in the next chapter.

5
Port and Trade Industrial Organization

Inevitably, the effectiveness of a port-integrated logistical system depends on the productivity and efficiency of ports, other components of the supply chain and trade industrial organization; this chapter will address these issues. First, port and supply chain productivity and efficiency will be investigated, because they serve as key performance measures of port-focal supply chain management. However, when compared to firm focal supply chains, port-focal supply chains contain ports that are not only production facilities, but are also transactional ones. Hence, it will continue to address the transaction cost economics of port, transport and international trade, pertinent to organizational dynamics of port-focal logistics and supply chains. The chapter concludes with a port efficiency and performance assessment, by investigating an empirical study on the performance benchmarking of global container ports.

5.1 Port and supply chain productivity and efficiency: production and transaction costs

Port and supply chain efficiency, a key performance measure of port-focal supply chain management, has been studied so far along the dimension of supply outputs, typically in terms of a single-output inventory system, such as the well studied newsvendor type (see Chapter 3). On the other hand, the forefront of the field of efficiency studies utilizes the economics of productivity and efficiency. A classical economic efficiency model that is suited for econometric analysis is the concept of production frontier – defined as the optimal performance level of a particular production system. For the sake of terminological clarity, one should note that production is broadly defined as transformation of necessary input into desirable output by means of manufacturing and service technologies and operations. In this regard, production frontier analysis is the central element in economic efficiency theory as pioneered by Arrow et al. (1961) and McFadden (1963). Indeed, frontier analysis has predominantly been developed upon a construction of quantity-cost optimization under the theory of the firm as a production function, where a firm maximizes profits with a cost-minimized choice of supply inputs. The construction of a production frontier is formulated in a heuristic framework, as follows:

$$(\text{growth}) \times (\text{production}) = \text{production frontier} \qquad (5.1)$$

where the laws of production technology are characterized by a 'production' function of regular (technical) input factors (e.g. capital, labor and material), which is interacted and modulated with a 'growth' function of irregular (non-technical) and exogenous factors (e.g. time and environment). Production frontier specifies the outputs of a firm, an industry or an entire economy as a function of all combinations of technical and non-technical inputs.

As noted in Chapter 3, according to the notion of transaction cost efficiency of SCM (Williamson, 2008), certain forms of transaction costs (e.g. the costs of governance, costs of coordination, and so on), as opposed to typical production costs (e.g. purchasing and inventory costs), can affect both allocative efficiency and governance efficiency of SCM. Regarding the former, transaction cost theory asserts that transaction costs may be mitigated by the exogenous adoption of proper organizational forms which, in turn, may generate reduced costs of asymmetric information, and thus reduced allocative inefficiency. As to governance efficiency, there exist other forms of transaction costs (e.g. costs of coordination and administrative control) as an exogenous source of operational inefficiency in addition to endogenous information asymmetry. Although supply chain governance is largely absent in the mainstream supply chain models, the impacts of transaction costs have been recognized by some supply chain researchers. For instance, Cachon and Lariviere (2005) found that the administrative cost of implementing a revenue-sharing scheme (a form of the transaction cost of governance), is a key (exogenous) factor that has limited practical adoption in revenue-sharing contracts. As far as can be ascertained from the literature, the efficiency of supply chains has been studied mainly along the dimension of output allocative efficiency (e.g. a single-output inventory system, such as the well studied newsvendor type), with little consideration of input allocative efficiency, and almost no attention given to governance efficiency.

5.1.1 The efficiency of logistics and supply chain management

Supply chain efficiency is concerned with an intrinsic measure of supply chain performance, consisting of three dimensions of measures, namely *operation, coordination* and *governance*. The operational efficiency of supply chains, a well developed branch of SCM, focuses on the science-of-choice approach to the management of supply chain operations (endogenous), and has been thoroughly covered in many existing SCM publications (e.g. Simchi-Levi, Kaminsky, and Simchi-Levi, 2000). Consequently, this chapter focuses on supply chain efficiency along the dimensions of coordination and governance. Supply chain coordination refers to the attainment of an incentive-induced equilibrium under a principal-agent type contract, with the underlying market as the ultimate source of revenue generation for supply chains as a whole. Supply chain governance refers to the administrative control and transactions under the supply contract terms.

5.1.2 Theory of the firm as a production function

Once again, *manufacturing*, and *production* in general, involves the transformation of inputs of resources and materials into outputs of goods and services. All in all, it still comes down to the basic competitive-equilibrium structure of Arrow and Debreu (1954). Consumers are perfectly informed about, and have preferences over, a group of available goods. Simultaneously, suppliers (producers) of the goods are sufficiently endowed with production-possibility sets (or technology-possibility sets). According to economics, all these agents are assumed to be price-takers, and are engaged in a competitive organization, i.e. the consumers maximize their welfare in terms

Demand Function $D(p)$:

$q = a - \beta p$

or,

$p = D^{-1}(q) = \overline{a} - \overline{\beta} q$

Cost Function :

$C(q) = F + \int_0^q C'(x)dx = F + aq + bq^2$

$C'(q) = a + 2bq$ (marginal cost)

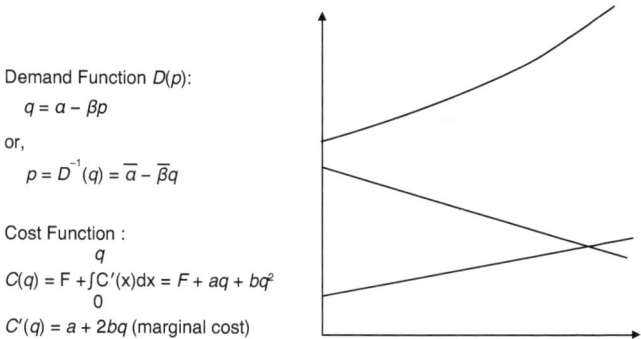

Figure 5.1   An example of demand and cost functions
Source: Authors.

of *demand function*, $D(p)$, which relates the desirable demand quantity D to market price p. Simultaneously, the suppliers maximize profits over their production possibilities in terms of *supply function* which relate the quantity of supply to the cost of inputs, i.e. the cost incurred in producing the supply, typically expressed in the form of a cost function $C(q)$ where q signifies the quantity of supply produced. Figure 5.1 illustrates a typical example of the demand and cost functions.

A *competitive equilibrium* is a set of prices with associated demands and supplies, denoted by the pair (p^*, q^*), such that the markets (one for each good) clear off (i.e. total demand does not exceed total supply). A key property of competitive equilibrium is that each good is sold at marginal cost, i.e. $p^* = c'(q^*)$ where $C'(q^*)$ is the marginal cost incurred for producing the equilibrium quantity of supply (Figure 5.2). This property is referred to in economics as Pareto optimality of competitive equilibrium, i.e. a fully informed 'social planner' (as so-called in the economics literature) could not find an alternative solution (i.e. allocation of goods) that would increase all the consumers' welfare. The intuitive reasoning goes like this. If the price of a good exceeds the marginal cost of producing it, the supplier would increase profit by expanding its production. On the other extreme, if the price of a good is smaller than its marginal cost, the supplier would contract its production, either partially or fully.

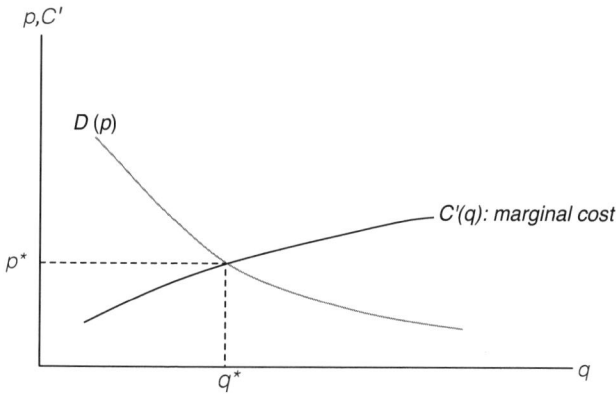

Figure 5.2 The economic equilibrium
Source: Authors.

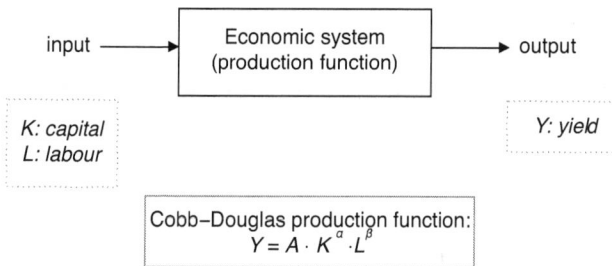

Figure 5.3 The economics of manufacturing (production function)
Source: Authors.

5.1.3 The production function

As to how supply is to be provided, economists often examine production and costs using the so-called *production function* which relates output supply to the required input demand (Figure 5.3).

In Figure 5.3, let *Y* be a maximized output of an economic system directly driven by the inputs, capital *K* and labor *L*, measured in physical units. A production function *f* is defined to be a real-valued function, expressed as:

$$Y = f(K, L) \tag{5.1}$$

Although we consider capital K and labor L to be both one-dimensional here, they can be multi-dimensional, such as $K = (K_1, K_2, \cdots, K_m)$ for multiple investment inputs and $L = (L_1, L_2, \cdots, L_n)$ for multiple variable inputs. A specific production function, for instance, is the Cobb-Douglas function, as follows:

$$Y = A \cdot K^{\alpha} \cdot L^{\beta} \tag{5.2}$$

where $\alpha + \beta$ gives the degree of homogeneity, and A is a coefficient of technological impact. Specifically, a production function is said to be:

- *Constant return to scale*, if $\alpha + \beta = 1$ which is termed *homogeneous of degree 1*.
- *Decreasing return to scale*, if $\alpha + \beta < 1$.
- *Increasing return to scale*, if $\alpha + \beta > 1$.

Let's consider the operational costs incurred in short-run production subject to a given production function. With any given desirable output level Q (e.g. annual demand) a firm must acquire necessary inputs with minimum costs (fixed cost on capital and variable cost on labor). In this case, let the cost be expressed as follows:

$$C = F + v \cdot Q,$$

where C is the total cost for acquiring the input, F is a fixed cost that is directly associated with the capital input K in a production function, and v is a variable cost in dollars per unit of output to be produced. Accordingly, the output level Q is governed by the production function, such as $Q = A \cdot K^{\alpha} L^{\beta}$. As the operational cost is mostly concerned with short-run production under a given capital setup, the capital input K is considered fixed, and therefore so is the capital cost F. In the short run, production incurs a process of seeking minimum cost in obtaining supply so as to meet demands.

5.1.4 Creative destruction and economic growth

Economists have long recognized the impact of science and technology on long term economic growth, and have based the impact on the theory that *technological change* is the most crucial element in economic growth (Romer, 1990). It is fair to argue that *economic growth theory* evolved largely around some controversial conjectures found in Schumpeter's (1942) influential work, entitled *Theory of Economic Development*. He theorized that:

> The hallmark of capitalism was...the dynamic evolutionary growth, which depended primarily upon innovations: new consumer goods, new methods of production and transportation, new markets, and new forms of industrial organization. (Schumpeter, 1942, p. 107)

Indeed, this theory inspired his enduring phrase *creative destruction*, first articulated in his masterpiece *Capitalism, Socialism, and Democracy* (Schumpeter, 1942). Technological change leads to superior products and services, while it simultaneously undermines the economic position of firms committed to older technologies. Schumpeter attributed the dramatic gains in economic growth (e.g. net income per capita) to *technological progress*. He insisted that technological progress (hereinafter called *technological change*), was a more plausible explanatory factor, and was both an endogenous and an exogenous factor that drives growth. He argued that:

> ...It is therefore quite wrong...to say...that capitalist enterprise was one, and technological progress a second, distinct factor in the observed development of output; they were essentially one and the same thing or...the former was the propelling force of the latter...(Schumpeter, 1942, p. 110)

Hence, one should not treat *technological change* in isolation from firms. Another of Schumpeter's more radical assertions goes against mainstream Western economic thought which extolled the virtues of competitive markets as the wellspring of prosperity. He argued that:

> In this respect, perfect competition is not only impossible, but (also) inferior, and has no title to being set up as a model of ideal efficiency. It is hence a mistake to base the theory of government regulation of industry on the principle that big business should

be made to work as the respective industry would work in perfect competition. (Schumpeter, 1942, p. 106)

Following his provocative assertions were the vivid and long lasting debates over a spectrum of economic issues. In the following, the authors have compiled two debated issues that are still highly relevant today:

- Equilibrium vs. non-equilibrium economy. First, let's clarify the terminology of *formal* theorizing versus *appreciative* theorizing. Here, *formal* theorizing is meant to be the deductive quantitative methods, such as optimization, as opposed to the inductive ones in *appreciative* theorizing. The typical formal theory of economics is based on the ideal goal of the economy: steady state equilibrium. However, Schumpeterian economists base their theory on the argument that equilibrium is only transient while non-equilibrium is prevalent, as characterized by the notion of *creative destruction*.
- Exogenous vs. endogenous factors. Earlier economic growth theory (e.g. Solow, 1957) set exogenous variables that measure *technological change*. A new technology is an external source for firms to generate increased productivity. After then, the growth theory is advanced to consider endogenous variables in technological innovation (Romer, 1990), such as the concept of human capital, which largely refers to the stock of technical knowledge. It has been shown that the rate of *technological change* is immensely affected by how individual firms invest in their human capital.

It is these ongoing debates that have brought about the *economic growth theory*. Economists view the total outputs of the economy as being due to various inputs (e.g. labor and capital) into the productive process. Within such a framework, the economic growth theory focuses on the *formal* theorization of the relationship between technological change and economic growth, taking inputs from *appreciative* theorization. The 'old' growth models were introduced in late 1950s, including Solow's (1956) classical work for which he received the Nobel Prize, among others including Abramovitz, 1956 and Kendrick, 1961. They attributed whatever portion of the measured growth of output could not be explained by the inputs to external technology change. A renewed interest in economic growth theory towards the end of the 1980s led to the development of 'new' growth models that differ from the 'old' ones by incorporating sources of endogenous growth (Romer 1990). To quantify the impacts of *techno-logical change* on economic growth, an *aggregate production function* is thus constructed as a time series as follows:

$$Y(t) = A(t) \cdot f(K, L) \tag{5.3}$$

With the special case of *neutral* technical change where the marginal impact of *technical change* is independent of that by K and L, the time-variant coefficient $A(t)$ gives the *cumulative* effect of technical change over time. In a quantified manner, the growth model (5.4) reveals important relationships between economic growth and technological changes, as follows:

- *Diminishing return on capital.* The higher the level of K, the less capital contributes to increasing Y.
- *Zero long-term growth.* As a result, capital accumulation becomes more and more difficult, eventually leading to zero growth in the long run.
- *Technology change.* The accumulation of capital, measured by capital per labor, is a direct consequence of technological change.

According to the classical production theory, the production function must satisfy the so-called regularity conditions which require the production function to be non-decreasing semi-continuous.[1] As the production function of a firm is never known in practice, it is suggested that the function be econometrically calibrated under the regularity conditions, and be estimated from empirical data, using either non-parametric (e.g. DEA: data envelopment analysis) or parametric (e.g. SFA: stochastic frontier analysis) methodologies.

5.1.5 Profit maximization

The profit of a firm is defined as the total revenue (TR) minus the total cost (TC) incurred in generating the output supply level of y and using a set of input demand $x = (x_1, \cdots, x_n)$, given output price p and input cost $w = (w_1, \cdots, w_n)$, as follows:

$$\pi = TR - TC = p \cdot y - w \cdot x^t \tag{5.4}$$

A profit-maximizing firm is to decide on an output level $\tilde{y}$ and input mix $\tilde{x}$, so as to maximize profits under a certain production technology ($y = f(x)$), given input and output prices. The (output) supply function under the profit maximization is denoted by:

$$\tilde{y} = \tilde{y}\ (p,\ w)$$

The corresponding (input) demand function is then written as:

$$\tilde{x}_i = \tilde{x}_i(p,\ w),\ i = 1, \cdots, n$$

Then, a profit function of a profit-maximizing firm can be defined as the maximum profit as a function of particular input and output prices, in the following form:

$$\tilde{\pi}(p,\ w) = p \cdot \tilde{y}(p,\ w) - w \cdot \tilde{x}^t(p,\ w)$$

where $\tilde{x}(p,\ w) = (\tilde{x}_1(p,\ w), \dots, \tilde{x}_n\ (p,\ w))$, and $w \cdot \tilde{x}^t(p,\ w) = \sum\limits_{i=1}^{n} w_i \tilde{x}_i\ (p,\ w)$.

Hotelling's Lemma

If the profit function $\tilde{\pi}(p,\ w)$ is differentiable, then the partial derivatives of the profit function provide the output supply equation and input demand equations as follows:

$$\frac{\partial \tilde{\pi}(p,\ w)}{\partial p} = \tilde{y}(p,\ w) \qquad \frac{\partial \tilde{\pi}(p,\ w)}{\partial w_i} = \tilde{x}_i(p,\ w), \quad i = 1, \dots, n$$

Under the regularity conditions of production function, the profit function, $\tilde{\pi}(p,\ w)$, has the following properties (see Beattie and Taylor 1985 for proofs):

1. $\tilde{\pi}(p,\ w)$ is nonnegative for nonnegative prices, i.e. $\tilde{\pi}(p,\ w) \geq 0$, for $p,\ w \geq 0$
2. $\tilde{\pi}(p,\ w)$ is non-decreasing in p
3. $\tilde{\pi}(p,\ w)$ is non-increasing in w
4. $\tilde{\pi}(p,\ w)$ is homogenous of degree one in all prices
5. $\dfrac{\partial \tilde{\pi}(p,\ w)}{\partial p}$ and $\dfrac{\partial \tilde{\pi}(p,\ w)}{\partial w_i}$ are homogeneous of degree zero in all prices

6. $\tilde{\pi}(p, w)$ is convex in all prices if the production function is strictly concave.

5.1.6 Cost minimization

When output supply is exogenously determined, input cost minimization with a given output level can be considered. Similarly, a cost function is defined as the minimum input cost of producing a particular output level y with given input prices w, as follows:

$$\tilde{c}(y; w) = w \cdot \tilde{x}^t(y; w)$$

where $w \cdot \tilde{x}^t = \sum_{i=1}^{n} w_i \tilde{x}_i(y; w)$. Then, we have the following Shephard's Lemma.

Shephard's Lemma

If the cost function $\tilde{c}(y, p, w)$ is differentiable, then the partial derivatives of the cost function give the input demand equations, given an output level y, as follows:

$$\frac{\partial \tilde{c}(p, w)}{\partial w_i} = \tilde{x}_i(y; w)$$

It is appropriate to mention that Hotelling's Lemma and Shephard's Lemma characterize the duality in production. It should be noted that a profit-maximizing firm is to determine input demand $\tilde{x}$; the cost function can provide an alternative way to obtain output supply and input demand functions. In the duality of production, the properties of the cost function can be obtained, as follows:

1. $\tilde{c}(y, w)$ is nonnegative for $w \geq 0$ and $y > 0$
2. $\tilde{c}(y, w)$ is non-decreasing in w
3. $\tilde{c}(y, w)$ is homogenous of degree one in all input prices
4. $\dfrac{\partial \tilde{c}(p, w)}{\partial w_i}$ are homogeneous of degree zero in all prices
5. $\tilde{c}(y, w)$ is weakly concave in input prices if the production function is strictly quasi-concave.

5.1.7   Total surplus and social welfare

The concept of total surplus is one of the most important foundation blocks in the theory of industrial organizations. For the sake of relevance to supply capacity, the authors elaborate on the total surplus through the contexts of supply contract under perfect information, i.e. both supplier and consumer have access to the same set of perfect information. Under this situation, an optimal supply contract $\{q^0, p^0\}$ satisfies the condition of competitive equilibrium, i.e. $p^0 = c'(q^0)$ (Figure 5.4).

5.1.8   Supplier surplus and consumer surplus

Suppose that the supplier starts fulfilling the contract, incurring a marginal cost of $c'(k)$ for producing the k-th item ($k = 1, 2, \cdots, q^0$). Since the contract value for each item produced is p^0, a surplus generated from producing an item k can be determined as $p^0 - c'(k) \geq 0$ for $k = 1, 2, \cdots, q^0$. Then, the total supplier surplus of the contract can be defined as:

$$\text{Supplier Surplus} = \int_0^{q^0} \left(p^0 - C'(x)\right) dx = \text{shaded area in Figure 5.5}$$

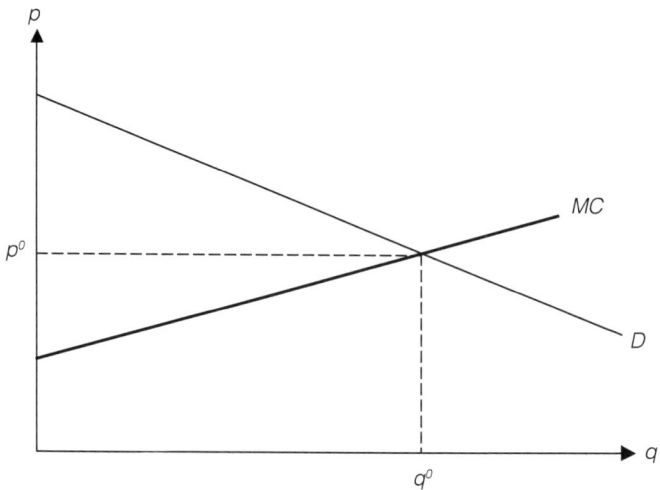

Figure 5.4   Supply contract under perfect information
Source: Authors.

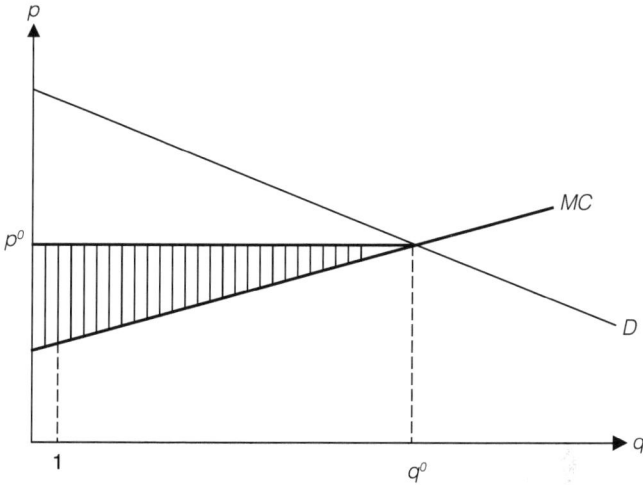

Figure 5.5 Total supplier surplus
Source: Authors.

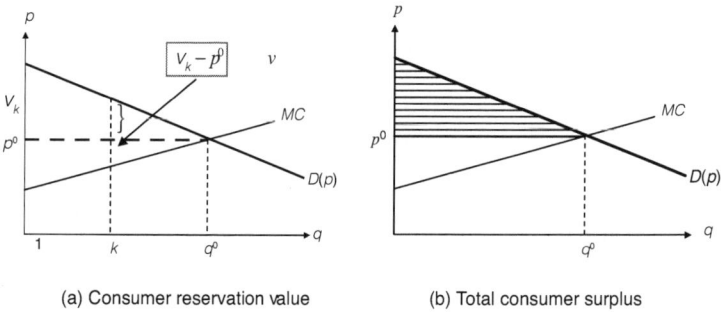

(a) Consumer reservation value　　　(b) Total consumer surplus

Figure 5.6 Reservation value and consumer surplus
Source: Authors.

The suppli er surplus represents a total value appreciable to the supplier by fulfilling the contract. Similarly on the consumer (buyer) side, the reservation value for the k-th purchase can be determined from the demand curve as v_k, as depicted in Figure 5.6a, and a surplus of $v_k - p^0$ can be appreciated by the k-th buyer. The total consumer surplus can be defined as:

$$\text{Consumer surplus} = \int_{p^0}^{\bar{p}} D(p)dp = \text{shaded area in Figure 5.6b.}$$

5.2   Transaction cost economics of port, transport and international trade

The global financial crisis of 2008 has taught the world an important lesson: that transport infrastructure interacts profoundly with international trade in terms of trade pattern and organization structure, far beyond the measures by traditional cost benefit analysis (CBA). A critical question that arises is how to characterize transport infrastructural effects on international trade.

5.2.1   Problems to address

What are the benefits of transport infrastructure and how can one measure such benefits? In the traditional CBA for transport infrastructure projects, the benefits are measured by quantifying both the direct impacts of a particular project on transport users (passengers and freight shippers) through reducing travel time and improving travel time reliability and the indirect effects through reducing various externalities of transportation (e.g. pollution, noise, accident and so on). In this regard, economic theories suggest that transport infrastructure has much wider effects than those being quantified in CBA. For instance, in the theory of New Economic Geography (NEG) (cf. Krugman, 1998), the reduction of transport cost from infrastructure investment would lead to agglomerations which increase the productivity of firms. Also, transport infrastructure can improve productivity through increasing labor supply by reducing commute costs (Cogan, 1981; Parry and Bento, 2001), and through reducing the inventory of firms by reducing transportation time (Shirley and Winston, 2004). Empirical studies on the causal effects of transport infrastructure on productivity normally estimate aggregate cost or production functions in which transport expenditure is one of the inputs.[2] The estimates on the rate of return of transport infrastructure from these studies have a wide range. There are empirical studies which investigate the broad effects of transport infrastructure on the economy through different channels.

Limao and Venables (2001) identify a positive relationship between infrastructure quality and trade; likewise Clark et al. (2004) found that the efficiency of ports facilitates international trade. On the other hand, Michaels (2008) identifies the effect of access to inter-state highways on the development of rural counties of the US, while Baum-Snow (2007) estimates the effect of highways on urban sprawl.

Generally speaking, traditional CBA tends to underestimate the impacts of transport infrastructure. Although existing studies attempt to identify the broad effects of transport infrastructure through different channels, there remains an urgent need for a framework to integrate the fragmented empirical evidence into a systematic measure of the long-run effects of transport infrastructure invest-ments. In the social savings approach of Fogel (1964), the impor-tance of railroads to the US economy is quantified by employing a counterfactual scenario: if the rail network were closed down for one year, what would be the GNP loss when the same volume of freight is carried to the same destinations by using alternative transport modes? The major criticism of the social savings approach is that such a counterfactual analysis is rather static (Williamson, 1974). As indicated by both theoretical analysis and empirical evidence, transport infrastructure can have profound impacts to the overall economy by affecting productivity and trade. Thus, it is far from being static and factual.

Recent advancements in theories of international trade, urban economics and industrial organization offer a valuable opportunity to revisit the problem of measuring the broad effects of transport infrastructure. For example, by using the New Ricardian Model developed by Eaton and Kortum (2002), Donaldson (2010) measures the gains in India due to rail transport in the nineteenth century. Additionally, based on the general gravity theory in Anderson and van Wincoop (2003), Duranton et al. (2011) measure the effects of interstate highways on trade and industry specialization in US cities. Unlike the reduced-form estimates in other empirical studies on the effects of transport on trade, the structural estimates in these two studies have clear interpretations on the mechanisms through which transport affects inter-regional trade. However, as in the standard trade theory, productivity in these two studies is taken as given, such that transport infrastructure is assumed to affect trade only through affecting trade costs. The assumption of exogenous productivity leads to the general underestimation of the impact of transport infrastructure.

Built upon recent theoretical advancements in urban economics and international trade, new frameworks are constructed to understand and measure the long-run impacts of transport infrastructure on the overall economy in a systematic way, with the following three component modules:

- Module 1: A theoretical framework to illustrate how transport infrastructure impacts trade through affecting both trade cost and productivity.
- Module 2: The ways to empirically verify the theoretical predictions.
- Module 3: Application methods to quantify the broad effects of transport infrastructure on the economy using compiled data.

5.2.2   Module 1: theoretical model

Objectives: In this module, a theoretical framework is built to show how transport infrastructure affects both the average level and the variation of firm productivity in a country or region. According to NEG, transport infrastructure affects firm productivity through the agglomeration effect. However, the most important role of transport infrastructure is to facilitate trade and there are influential studies in the trade filed on the productive efficiency gains from open to trade. A good summary on these studies can be found in Melitz and Trefler (2012). In this chapter, the theoretical framework will integrate two strands found in the literature. As such, under the framework, transportation infrastructure affects firm productivity via two channels: agglomeration effect and selection/sorting effect. The intuition of the sorting effect is that lower trade costs as a result of transport infrastructure investment increase market competition and thus increase the productivity threshold of firm entry; given a high quality of transport infrastructure, only firms with high productivity can enter. The main expected predictions from the theory, which contribute to the current literature, are that transport infrastructure can not only increase the average productivity level but also reduce the variation in firm productivity.

Context: Consider a world economy with multiple countries and use $c \in \{1, \cdots, C\}$ to index the countries. There are two cities, indexed by 1 and 2, in each country. Without loss of generality, cities 1 and 2 of each country are considered as the port city and inland city, respectively. There are two types of transport infrastructure: 'inland' infrastructure connecting the two cities of a country (highways or railways) and 'export' infrastructure connecting the port cities of different countries (seaports or airports). Given this setup, the broad effects of improving a country's transport infrastructure, including both the inland and export infrastructures, will be investigated. There is a continuum of consumers with mass M_c in country c and consumers in all the countries have an identical preference on a variety of products. The utility of a representative consumer is given by the standard constant elasticity of substitution (CES) utility function, as follows:

$$U = \left\{ \int_0^N q(j)^{\frac{(\varepsilon-1)}{\varepsilon}} dj \right\}^{\frac{\varepsilon}{(\varepsilon-1)}}$$

where $q(j)$ is the consumption on the variety $j \in [0, N]$ and ε is the elasticity of substitution between any two varieties. Although individuals are homogeneous in their preferences, they are different in their talent, which determines the productivity level of individuals, and in their luck, which determines the entry cost to the individuals to become entrepreneurs that produce the product varieties. Therefore, an individual can be indexed by the ordered pair (ρ, f), where ρ denotes talent and f denotes luck; ρ and f are random draws from their own distribution functions, which are independent of each other and are denoted by $G_\rho(\cdot)$ and $G_f(\cdot)$ respectively. The timing of events in the model is as follows. First, the talent of each individual is realized as a random draw from $G_\rho(\cdot)$. Second, by knowing his/her own talent, an individual in a country chooses one of the two cities to in which to live. Third, after residing in a city, the luck of each individual is realized as a random draw from $G_f(\cdot)$. Fourth, given the chosen location and by knowing both his/her own talent and luck, an individual chooses to become either an entrepreneur or a worker. Finally, product varieties are produced by the entrepreneurs and are traded across cities and among countries. Trade costs take the standard 'iceberg' form, i.e. in order to transport one unit of product j from origin o to destination d, the quantity τ_{od}^{j} should be produced and transported, where $\tau_{od}^{j} \geq 1$. Mobility is limited in this model, though, as individuals cannot move to another city after they choose their occupations; also, people cannot migrate across countries.

Market structure in this model is the monopolistic competition as in Dixit and Stiglitz (1977), such that an entrepreneur produces a product variety. In this case, labor is the only input of the production. If a firm in a country is set up by an entrepreneur indexed by (ρ_e, f_e), then the setup cost of the firm is f_e and the firm requires $\phi(\rho_e)$ efficiency units of labor to produce a unit of a variety; $\phi(\cdot)$ is a strictly decreasing function of talent. A worker indexed by (ρ_w, f_w) provides $h(\rho_w)$ efficiency units of labor; $h(\cdot)$ is also a strictly decreasing function of talent. Under these conditions, the labor demand of the firm, which produces product variety j, is given by

$$l_j = q(j) \times \phi(\rho_e) \tag{5.6}$$

where l_j is the efficiency units of labor demanded by the entrepreneur. In this regard, the equilibrium concept of the model is given by:

Definition 1. *An equilibrium is a collection of continuous allocation and price functions such that (i) given realization of talent, each individual in each country optimally chooses a city to live; (ii) given realizations of both talent and luck, each individual in each city optimally chooses an occupation; (iii) given prices of product varieties, each entrepreneur in each city maximizes her profits and markets of product varieties clear; (iv) given wage rates of cities, labor markets in the cities clear; and (v) population adding-up constraints are satisfied.*

Equilibrium analysis is typically conducted for three scenarios. First, there are not both export and inland infrastructures. Second, there is export but no inland infrastructure. Finally, there are both export and inland transport infrastructures. The characterizations show the productivity gains, i.e. increase in average productivity level and reduction in firm productivity variation, from investing in transport infrastructure.

5.2.3 Module 2: empirical models to test the theoretical predictions

Objectives: This module develops empirical models to test the theoretical predictions of module 1. The Ricardian trade model predicts that countries should produce and export relatively more in industries in which they are relatively more productive. The seminal work in Eaton and Kortum (2002) formalizes Ricardo's ideas and empirically quantifies the impacts of relative productivity on trade. Constinot et al. (2012) generalize Eaton and Kortum's model and empirically explore Ricardo's ideas. The empirical approach developed in this module is to test the theoretical predictions from the previous module; the approach is based on the basic framework developed by Eaton and Kortum. Intuitively, since transport infrastructure affects the productivity of a country or region, such impacts should affect the trade between the country and other countries. As such, by looking at the trade data, it is possible to quantify the productivity gains from transport infrastructure investments.

Context: The generalization of Eaton and Kortum's model in Constinot et al. (2012) is as follows. Let k index goods in a particular country and other countries are indexed by i. Each industry consists of an infinite number of varieties, which are indexed by ω. The term $z_i^k(\omega)$, which denotes of productivity of country i in producing variety ω of good k, is the number of units of the ωth variety of good k that can be produced by one unit of labor in country i. The key point in Eaton and Kortum is to model productivity as a random draw from the following extreme value distribution parameterized by z_i^k and θ, which can be expressed as:

$$F_i^k(z) = \exp(-(z/z_i^k)^\theta), \quad \text{for all } z > 0 \tag{5.7}$$

where z_i^k represents the fundamental productivity of country i in industry k and θ represents the intra-industry productivity difference, which is assumed to be common across countries and industries in the trade literature. In the New Ricardian trade theory, the two parameters play and important role in determining the goods that countries trade and the gains of countries from open to trade. The theory developed in module 1 predicts that both z_i^k and θ are affected by a country's transport infrastructure. In the empirical analysis, equation (3) is modified so as to let θ vary across both countries and industries and the modified parameter is denoted by θ_i^k. The research questions in this module aim to specify an empirical model that identifies how z_i^k and θ depend on a country's transport infrastructure. The findings to these questions will make significant contributions to the fields of both trade and transport.

An important implication of Eaton and Kortum's model is that bilateral trade flows take the gravity equation form, which has been widely adopted by empirical studies in the field of international trade. By doing so, one can specify an empirical gravity type trade flow equation, which is guided by the New Ricardian trade theory and is parameterized by a set of unknown parameters including z_i^k and θ_i^k. The empirical specification allows the two productivity parameters to vary across countries and part of the cross-country variation is attributed to the quality of transport infrastructure. The empirical specification also allows the two productivity parameters to vary across industries such that the impacts of transport infrastructure on productivity can be different across industries. Nevertheless, the identification to the specified empirical trade flow equation may face substantial challenges, such as omitted country effects and simultaneity of transport infrastructure investments and trade flows. Hence, researchers can examine these identification challenges carefully and explore different identification strategies which are guided by both econometrics and economics theory.

5.2.4   Module 3: quantifying the gains of transport infrastructures

Objectives: Armed with the theoretical foundation built in module 1 and the empirical model developed in module 2, module 3 quantifies the impact of transport infrastructure investments on trade. Previous studies on this topic by Donaldson (2010) and Duranton et al. (2011) quantify the impact of transport infrastructure on trade through the channel of reducing trade costs. This chapter will contribute to the literature by quantifying the impact of transport infrastructure on trade through both reducing trade costs and changing firm productivity through compiling data from the People's Republic of China (hereinafter called 'China'). Based on the compiled data, the researchers can conduct the following analyzes:

- Estimate the two productivity parameters in the New Ricardian trade model based on the empirical model in module 2;
- Quantify the total effect of transport infrastructure on both China's interregional trade and international trade between China and other countries by using the estimated parameters. Also, the authors will decompose the total effect into the individual effect from the two channels; and
- Estimate the welfare effects of transport infrastructure investment through the income effects.

Context: In the past three decades, China has experienced rapid growth in both exports and imports. Today, China is one of the world's largest trading nations. The rapid growth of trade in China is accompanied by a huge investment in transport infrastructure including highways, railways, seaports and airports. Such a large variation in both trade and transport infrastructure over time provides a rare opportunity to investigate empirically the effect of transport infrastructure on trade. In this module, the authors first ensemble a comprehensive dataset which records the following information:

- Trade volume and components among different regions (provinces) in China over time;
- Trade volume and components between different regions of China and other countries over time;
- Stock of different types of transport infrastructure – seaport, airport, highway and railway, in different regions of China over time; and
- Other economic variables such as population, gross domestic product (GDP), and industry components of different regions of China over time.

Given the data set, researchers can conduct extensive empirical analyzes which are guided by theories and empirical models developed in previous modules. Specifically, this module answers the following questions from the empirical analyzes:

- What is the effect of transport infrastructure investment on overall productivity level and firm productivity distribution of different industries in different regions of China? The empirical models to answer this question are the ones developed in module 2;

- What is the effect of transport infrastructure on population growth and industry component in different regions of China? The analysis is this issue tests the predicted agglomeration effect from transport infrastructure in module 1;
- What is the effect of agglomeration on productivity level of different regions? The answer to this question, in combination with the answer to the first question, led to a further question: How does transport infrastructure affect productivity through the two channels: agglomeration and the sorting mechanism?
- What is the impact of productivity change on trade? The authors will run counterfactual simulations by using the New Ricardian trade model to show what the trade patterns would be if the transport infrastructure investments did not occur; and
- What are the welfare effects of the impacts of transport infrastructure on trade? Following Donaldson (2010), the authors will use the income effects of trade to uncover the overall welfare effects of transport infrastructure that affects trade patterns.

5.3   Port efficiency and performance assessment

With port being a core trade facility and the infrastructure of port-focal logistics, port efficiency interactively represents integrated performances of both logistics and trade. In the following, an illustrative example on benchmarking container port performance will be given.

5.3.1   Container port performance benchmarking

Since the mid-1970s, most general cargos, traditionally carried by the break-bulk method, have been carried in containers (see Chapter 2), and the container port industry has become a very important link in the international trade network. Consequently, as mentioned earlier, supply chain logistics has evolved from being firm focal (i.e. inbound and outbound with a company at its center) to port-focal (i.e. in and out with a seaport/airport as the center). Integrating with port logistics which is broadly defined as transport logistics of seaports, airports and dry ports/inland terminals (see Chapter 3), supply chain logistics has become the artery of international trade and global supply chain operations, especially for the contract-manufacturing based retail business, e.g. WalMart, Ikea, Target, and so on. A port-focal supply chain, especially of a WalMart-type business, operates on a selective network of heterogeneous contract manufacturers (e.g. of different industries and products), together with a direct distribution network of supermarkets of its own band (see Chapter 3 for a more detailed discussion); while on the other hand, a firm based supply chain (e.g. a manufacturer-supplier, or manufacturer-distributor type) is homogeneous in the sense that it consists of firms in the same industry with each firm as a production function. An example of port-focal supply chain logistics engaged in a port-operator logistics (POL) system with the port as a governance structure is illustrated in Figure 5.7.

Port production: Port production is of a logistical type, of which production output is measured in terms of volume of containers, and production input consists of regular 'capital inputs,' which are divided into three categories of cargo handling equipment, terminal infrastructure, and storage facilities (e.g. cargo handling capacity, cargo storage capacity and fixed capital). In addition, port inputs also include several *individual characteristics* and *environmental variables* (i.e. degenerative inputs), such as the number of operators in port, the level of managerial coordination, GDP, import (IMP) and export (EXP). That is, 'capital inputs' are regular and endogenous in terms of port technology as a production function, and individual characteristics are degenerative and exogenous in terms of individual

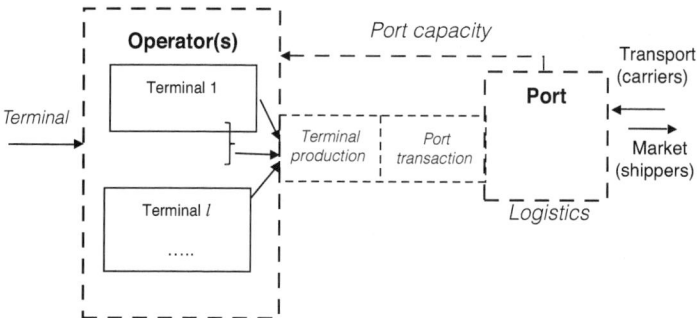

Figure 5.7 A port-operator logistics system
Source: Authors.

system heterogeneity in port production. In sum, a port production function is expressed in terms of output volume (y) as a function of regular input (x), as follows:

$$y = f(x) = A(\tau) \cdot g(x) \tag{5.13}$$

where $A(\tau)$ represents the growth/environment coefficient as a function of (exogenous) system parameters τ.

Efficiency measurement: This sub-section presents a brief review of the classical productivity and efficiency theory, as a necessary methodological foundation for efficiency analysis of supply chain management. The production efficiency (PE) of a firm, as a production function in terms of transforming inputs into desirable output with a certain production technology of either a manufacturing or a service type, is broadly defined as a ratio of the actual productivity (AP) measure to the relevant frontier productivity (FP) measure. That is:

$$production\ efficiency\ (PE)\ =\ \frac{actual\ productivity\ (AP)}{frontier\ productivity\ (FP)}$$

According to the classical efficiency theory, production efficiency is measured along two dimensions: (i) *technical efficiency*, which measures the firm's performance in terms of maximizing output from a given set of inputs; and (ii) *allocative efficiency*, which measures the firm's performance in terms of allocation of inputs, outputs and between the two, given respective prices and a production technology, i.e. the combination of the two component measures provides a measure of total *economic efficiency*, in a generic formulation of:

$$economic\ efficiency\ =\ (allocative\ efficiency) + (technical\ efficiency)$$

The classical efficiency measurement is gauged upon two measurement constructions, namely an input-based efficiency measurement, based on an effective selection of input mix to produce a certain mix of outputs, and an output-based efficiency measurement, which is based on an effective attainment of output quantities with a given set of inputs.

5.3.2 Classical frontier efficiency analysis

Production frontier analysis: A classical economic efficiency model that is suited for econometric analysis is the well-known concept of production frontier, defined as the optimal performance level of a production system. Production frontier analysis is the central element in economic efficiency theory, as pioneered by Arrow et al. (1961) and McFadden (1963). Frontier analysis has predominantly been developed upon a construction of quantity-cost optimization under the theory of the firm as a production function, where firms maximize profit with a cost-minimized choice of supply input. The construction of production frontier is formulated in a heuristic framework as follows:

(growth) × (production) = production frontier

where the laws of production technology are characterized by a production function of regular (technical) input factors (e.g. capital, labor, material), which is interacted and modulated with a 'growth' function of irregular (non-technical) exogenous parameters. Production frontier specifies the output of a firm, an industry, or an entire economy as a function of all combinations of technical and non-technical inputs.

Preserving the regularity conditions, the technical input factors must be 'regular' in terms of: (i) *homogeneity*, i.e. the production function is homogeneous with regard to technical input factors as independent variables; and (ii) *cost linearity*, i.e. the total input cost can be completely measured as an inner product of unit costs and quantities supplied. The production frontier of a cost-minimizing firm is defined as the functional output $\tilde{y} = A \cdot g(\tilde{x})$, under the theory of the firm as a production function, where A is the (exogenous) growth coefficient, and $g(\tilde{x})$ represents the choice of cost-minimized inputs, $\tilde{x}$ which can be solved from the following cost minimization problem with given input prices (vector) w:

$$\begin{cases} \min_{x} w \cdot x^t &= \sum_{i=1}^{n} w_i x_i \\ s.t. & x \in L(y) \equiv \{x : f(x) = A \cdot g(x) \ge y, \quad \text{for } y \ge 0\} \\ & \text{given}: y \ge 0, \quad w \ge 0 \end{cases} \qquad (5.14)$$

Input-allocative technical frontier efficiency: Then, an input-allocative technical efficiency of the firm as a homogeneous function of the 'regular' technical inputs is measured against its frontier $\tilde{y} = A \cdot g(\tilde{x})$ as an optimal benchmark. Let $\hat{y}$ denote the actual production output with frontier input $\tilde{x}$, with an inefficiency factor denoted by $\Delta \geq 0$, expressed in the following format:

$$\hat{y} = f(\tilde{x}) \cdot e^{-\Delta} \leq f(\tilde{x}) \tag{5.15}$$

That is, only if the firm is efficient (i.e. $\Delta = 0$), does the firm reach the frontier $\tilde{y} = f(\tilde{x}) = A \cdot g(x)$. Given the same cost-minimizing input $\tilde{x}$, the input-allocative technical frontier efficiency (TFE_{input}) can be calculated as:

$$TFE_{input} = \frac{\hat{y}}{\tilde{y}} = e^{-\Delta} \tag{5.16}$$

Conversely, the technical inefficiency is given as $1 - e^{-\Delta}$.

Remarks: input and output based efficiency measures: One should note the following remarks on frontier efficiency measures that the authors have defined:

1. Since the production function is never known in practice and neither is the inefficiency term, practical assessment of the frontier efficiency of a firm relies on the estimation of production (frontier) function of a firm, together with the inefficiency term, from sample data. The major methods for frontier function estimation, and thus efficiency assessment, include a non-parametrical (e.g. DEA) and a parametrical (e.g. SFA) approach;

2. Input-allocative versus input-based efficiency measures. If input prices (w_i's) are known and $\tilde{x}$ can be determined from frontier model (5.14), the TEF_{input} given in (5.16) is of an input allocative type. Otherwise, if the input prices are unknown, the technical frontier efficiency as given in (5.16) is called an input based type (as opposed to an input allocative type), of which an input-based frontier output ($\tilde{y}$) can be calculated as maximum outputs achievable for a given set of input combinations, without using the frontier model (5.14) which requires input prices;

3. In theory, production output y can be multiple dimensions. For example, in a port, there can be two outputs, one of containers (y_1) and the other of bulk cargos (y_2). In practice, the outputs are measured separately along each of the output dimensions. No particular composite measurement of multiple outputs has been practically adopted;

4. Output-allocative efficiency measures. By production duality, an output-allocative efficiency can be constructed similar to revenue-maximizing output mix ($\tilde{y}$), subject to a given constraint region of inputs ($x \in X$), given respective output prices and the production technology. Similarly, if the output prices are unknown, an output-based efficiency is measured based on the maximum output frontier achievable from a given set of input combinations, without using output prices; and

5. The allocative efficiency measures above are devised from either a cost minimization construction or a revenue-minimization construction, but not from a profit maximization construction which requires both input and output prices.

Frontier efficiency assessment: As mentioned earlier, the production function is never known in practice, and thus the function and the derived efficiency measures need to be estimated from empirical data. In this sub-section, two major econometrical frontier efficiency models, namely SFA and DEA, will be elaborated through the prototype example of global container port performance benchmarking.

Case study: benchmarking the performance of container ports: Although the efficiency of container ports has drawn much attention from both academic research and governmental policy agendas, it is still lacking a rigorous modeling framework that can take the intrinsic characteristics of this industry into account in evaluating efficiency. There are two alternative approaches applied to the empirical assessment of technical efficiencies of container ports, namely (non-parametric) DEA and (parametric) SFA. However, two intrinsic characteristics of the port industry: the individual heterogeneity in production technology and the time-varying nature of technical efficiency, generally have been ignored in these research studies. In sum, several fundamental questions regarding port logistics efficiency, and therefore supply chain logistics, remain open.

Individual ports face different natural conditions and business environments which are largely uncontrollable for port management. Given the same inputs but under different outside factors, the possible maximal outputs of two ports are likely to differ. Such a difference in outputs should be interpreted as technology heterogeneity rather than efficiency differences because both of the two operators are in fact using their inputs in the best way. In this regard, DEA cannot deal with the technology heterogeneity, and the heterogeneity has not been taken into account in the aforementioned SFA studies on port efficiency. In sum, a container port herein is engaged in port production to generate necessary container throughput in TEUs, with a certain production technology as a function of regular technical inputs and individual characteristics (Table 5.1).

Data collection and analysis: The data used in this chapter cover operators from the world's top 100 container ports from 1997 to 2004 (ranked in 2005). Restricted to container operators, the outputs can be reasonably measured as the containers (in TEU) handled within a particular year. This output measure is a standard one used in the industry and has been used by most academic studies in the field. The capital inputs used to handle containers are classified into the following categories: cargo handling equipment, terminal infrastructure, and storage facilities. Cargo handling equipment is further divided into two types: those at the quay side, such as quay cranes and ship shore cranes, loading and discharging containers to and from ships; and those at the yard, such as forklifts and yard cranes, moving containers at the storage area. The terminal infrastructure is measured on the basis of the number of berths, the length of quay line, and the terminal area. As for the storage facilities, the inputs are measured by using the storage capacity and the number of electric reefer points. The data on the output and capital inputs are compiled from the Containerisation International Yearbooks 1997–2004 published by Lloyd's Maritime Intelligence Unit (MIU) (now Lloyd's List Intelligence). Unfortunately, it is difficult to collect credible data on the labor inputs of the terminal operators. The assumption that labor inputs are ignorable is that the ratio between capital and labor inputs varies little across operators. Finally, measures of port/terminal characteristics include water depth, the number of calling liners, and the number of operators and terminals in a port as proxies of possible missing inputs causing technology heterogeneity at the port/terminal level. These variables are also compiled from the Containerisation International Yearbooks 1997–2004 (Table 5.1).

SFA model: For a cost-minimizing port $k(i = 1, \cdots, m)$, a stochastic frontier model is constructed by taking a logarithm of the frontier production function given in (5.15), i.e. $Y = \ln(\hat{y})$ with $\hat{y}$ given by (5.15), as follows:

$$Y^k = \alpha_k + B \cdot X^k - \Delta_k + \varepsilon_k \quad , \quad k = 1, \cdots, m; \tag{5.17}$$

where X^k represents the technical input mix used by port k; $\alpha_k = \ln(A_k)$ is the logarithm of individual growth coefficient; $B = (B_1, \cdots, B_n)$ is a frontier slope vector; Δ_k is the non-negative technical inefficient term; and ε_i is an error term which accounts for random measurement error and other random factors. Thus, $\alpha_k + B \cdot X^k$ gives the logarithm of the technical frontier function of port k, that is,

$$\hat{y}^k = f(X^k) = A_k e^{B \cdot X^k} = e^{\alpha_k + BX^k} \tag{5.18}$$

By the terms of technical inefficiency (Δ_k) and random measurement error ε_k, which are only measurable statistically, the model (5.17), called an econometrical stochastic frontier model (cf. Aigner et al., 1977; Coelli et al., 1998), can be econometrically calibrated with sample production data without information on input prices. Assuming a cost-minimization port as a production function, the technical efficiency of each port (k) can be estimated as follows:

$$TFE^k = \frac{f(X^k)e^{-\Delta_k}}{f(X^k)} = e^{-\Delta_k} \tag{5.19}$$

A geometrical description of the stochastic frontier model is illustrated in Figure 5.8.

Table 5.1 The input and output statistics of global container ports

Variables	Mean	Std. Dev.	Minimum	Maximum
A. Terminal Output				
TEU: Container Throughput in TEUs (000's)	936.4	1,741.7	4.6	20,600
B. Terminal Inputs				
1. Cargo Handling Equipment:				
Cargo handling capacity at quay in tonnage [a]	385.0	470.7	23.9	5,416.2
Cargo handling capacity at yard in tonnage [b]	5,116.5	7,060.9	38.6	62,731.8
2. Terminal Infrastructure:				
Number of berths	5.1	5.2	1	37
Length of quay line in meters	1,361.3	1,181.6	200	9,000
Terminal area in squared meters (000's)	604.9	844.6	7.7	8,092
3. Storage Facilities:				
Storage capacity in number of TEUs (000's)	23.2	72.4	0.6	1,200
Number of electric reefer points	480.6	539.7	4	3,768
C. Individual Characteristics				
1. Terminal and port level:				
EDI (in fraction of total sample)	0.3			
Depth of water in meters	13.2	3.5	4.5	32.0
Number of liners calling at the terminal	16.2	14.5	1	114
Number of operators in port	3.7	2.6	1	10
Number of terminals in port	6.8	6.2	1	31
2. Operator group dummies (in fraction of total sample):				
Global Carrier	0.09			
Global Stevedore	0.15			
Other: not belong to any of above groups	0.76			

3. Country Characteristics:

GDP in current US dollars (billion) [c]	2,240	3,270	12,500
Goods exports in US dollars (billion) [c]	271	249	972
Goods imports in US dollars (billion) [c]	308	365	1,670
GDP per capita in current US dollars [c]	18,654.9	12,367.8	37,651

4. Continental Distribution (in fraction of total sample):

Asia	0.37
Europe	0.27
North America	0.17
Latin America	0.06
Oceania	0.09
Africa	0.04
Period	1997–2004
Number of Countries	39
Number of Ports	78
Number of Terminal Operators	141
Number of Terminals	397
Number of Observations	597

[a] An aggregate of (1) Quay cranes and (2) Ship shore container gantries.

[b] An aggregate of (1) Gantry cranes, (2) Yard cranes, (3) Yard gantries, (4) Reachstackers, (5) Yard tractors, (6) Yard chassis trailers, (7) Forklifts, (8) Straddle carriers, (9) Container lifters, and (10) Mobile cranes.

[c] The country data can be found at the World Bank website: http://devdata.worldbank.org/dataonline/old-default.htm

Source: Authors.

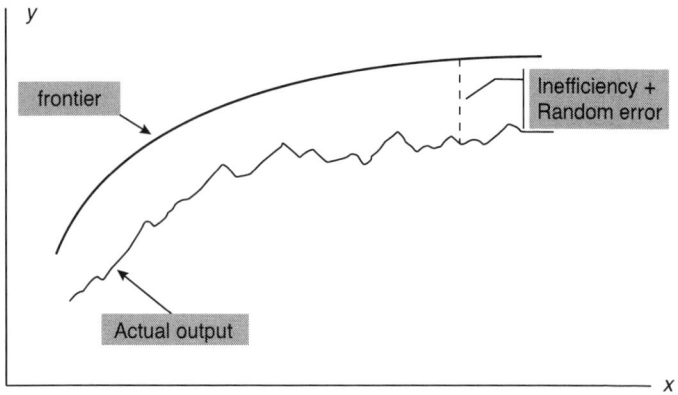

Figure 5.8   The stochastic frontier model: technical inefficiency
Source: Authors.

The estimation of technical frontier efficiency: Assuming that each port k is cost minimizing with input prices unknown, the stochastic frontier model is used to estimate the technical frontier $e^{\alpha_k + B \cdot X^k}$ in terms of intercept coefficient α_k and technical frontier slope vector B. Together with the technical inefficiency term Δ_k, all the coefficient terms of the stochastic frontier model need to be estimated from sample data of each port k.

   To develop an estimation method of the stochastic frontier model, we let $\hat{a}_k = \alpha_k - \Delta_k$, and rewrite it in a reduced form as follows,

$$Y^k = \hat{a}_k + B \cdot X^k + \varepsilon_k$$

$$(5.20)$$

Given a set of sample data in the format of (Y_t^k, X_t^k) for period t, for $t = 1, 2, \cdots, T$, the reduced stochastic frontier model (5.20) can be estimated in terms of $\hat{a}_k = \alpha_k - \Delta_k$ and B. Now, the key to further assessing the technical efficiency (TFE^k) relies on differentiation of the inefficiency term Δ_k from the intercept term α_k. To this end, we consider the individual technical characteristics of Table 5.1, denoted by vector τ^k for port k. Using the term $A(\tau)$ as defined in the production frontier function (5.13), we construct a second level of an estimation equation on the intercept term α_k of the stochastic frontier model (5.17), as follows:

$$\alpha_k = \beta_k + \Theta \cdot \tau^k + \nu_k \tag{5.21}$$

where β_k is an intercept term; Θ represents the slope vector of technical characteristics; τ_k is the vector of technical characteristics of port k (e.g. time, number of terminals, and so on as listed in category C of Table 5.1); and ν_k is another error term. According to the stochastic frontier model (5.17) and production frontier function (5.13), the inefficiency Δ_k is independent of technical characteristics τ^k. Thus, an econometrical SFA model can be constructed by combining both estimation systems of (5.17) and (5.21):

$$\begin{cases} Y_{kt} = \alpha_{kt} + B \cdot X_{kt} - \Delta_{kt} + \varepsilon_{kt} \\ \alpha_{kt} = \beta_{kt} + \Theta \cdot \tau_k + \nu_{kt} \\ k = 1, \cdots, m; \quad t = 1, \cdots, T \end{cases} \tag{5.22}$$

Case study: benchmarking of the performance of global container ports: This sub-section illustrates the results of applying the SFA model to the case study of global container ports performance benchmarking, using the data set as highlighted in Table 5.1. The port's production frontier is assumed to be a Cobb-Douglas type.[3]

Port production frontier estimation: The results of econometrical SFA estimation of the port frontier are presented in Table 5.2.

The first column of Table 5.2 presents the estimation results by the basic SFA model (5.22). The second column presents the estimation results by a reduced SFA model, by not controlling the individual characteristics τ^k. Finally, the third column gives the estimated results by the model without controlling the unobservable individual heterogeneity (i.e. $\nu_k = 0$).

Efficiency benchmarking results by SFA: The estimated results can be found in Table 5.3.

DEA model: Alternatively, the production frontier can be estimated by the non-parametric DEA, which only requires a non-parametric form of the production function.[4]

Table 5.2 The stochastic frontier analytical (SFA) estimation of port production frontier

Variable	Heterogeneous model	Conventional model
1. Log inputs		
Quay superstructure (β_1)	0.1815 (0.0725)	0.2142 (0.0581)
Yard equipment (β_2)	0.0200 (0.0361)	0.0527 (0.0315)
Berth number (β_3)	0.0953 (0.0509)	0.0290 (0.0576)
Quay length (β_4)	0.0802 (0.0507)	0.1228 (0.0589)
Terminal area (β_5)	0.0268 (0.0582)	0.0912 (0.0444)
Storage capacity (β_6)	0.0087 (0.0229)	−0.0406 (0.0274)
Reefer points (β_7)	0.1400 (0.0355)	0.2125 (0.0320)
2. Individual intercept		
Constant (θ_0)	−0.7063 (0.3121)	0.8267 (0.3495)
2.1 Port characteristics		
Water depth (θ_1)	0.4184 (0.2770)	0.6644 (0.2626)
Ship calls (θ_2)	0.1322 (0.0369)	0.1550 (0.0399)
Number of operators (θ_3)	−0.0520 (0.0985)	−0.3789 (0.0761)
Number of terminals (θ_4)	−0.0219 (0.0886)	0.1992 (0.1039)
2.2 Country characteristics		
GDP (θ_5)	−0.3815 (0.0872)	0.2437 (0.0972)
Goods exports (θ_6)	0.0660 (0.1024)	−0.1715 (0.0719)
Goods imports (θ_7)	0.3088 (0.1114)	−0.0023 (0.0697)
GDP per capita (θ_8)	−0.7931 (0.2498)	−0.4889 (0.1123)
2.3 Operator Group		
Carrier (θ_9)	0.3594 (0.2344)	0.3780 (0.1626)
Stevedore (θ_{10})	0.4538 (0.1874)	1.6053 (0.3623)
2.4 Time trend		
Time (r_1)	−0.0719 (0.0870)	−0.5350 (0.2197)
Time squared (r_2)	0.1031 (0.0430)	0.1553 (0.0996)
3. Variance of the constant (σ_v^2)	0.4437 (0.0650)	
4. Inefficiency parameters		
Coeff. of Constant (g_1)	−1.8584 (0.3797)	0.3517 (0.1554)
Coeff. of Time (g_2)	−0.5967 (0.4785)	−0.4790 (0.1711)
Coeff. of Time squared (g_3)	−1.1674 (0.2398)	−0.0812 (0.0913)
Coeff. of Carrier (g_4)	−0.3961 (0.4313)	0.7357 (0.3945)
Coeff. of Stevedore (g_5)	−1.0024 (1.0824)	−0.1368 (0.2601)
Variances (Σ_{11})	2.2743 (0.7468)	0.3327 (0.0838)
Variances (Σ_{22})	2.0274 (0.8271)	0.2680 (0.0688)
Variances (Σ_{33})	1.4244 (0.4569)	0.1736 (0.0386)
Variances (Σ_{44})	1.5650 (0.9784)	0.7158 (0.4536)
Variances (Σ_{55})	1.9605 (1.1155)	0.5996 (0.2482)
5. Other parameters		
Variance of noise (σ_ε^2)	0.0321 (0.0029)	0.0371 (0.0034)

Remarks: The numbers in parentheses are the posterior standard deviations. Also, all the input and output variables are normalized with respect to their sample means before taking log.

Source: Authors.

Consider that a constant-returns-to-scale (CRS) port k ($k = 1, \cdots, m$) produces an l-dimension output vector $y^k = (y_j^k)_{j=1}^l$, using an n-dimension input (vector) $x^k = (x_i^k)_{i=1}^n$. The basic DEA model is formulated as a mathematical programming problem: For each port k, find the weights $u^k = (u_j^k)_{j=1}^l$ and $v^k = (v_i^k)_{i=1}^n$ that solve the following linear programming:

$$\begin{cases} \max_{u^k, v^k} \quad u^k \cdot y^k = \sum_{j=1}^l u_j^k y_j^k \\ s.t. \quad v^k \cdot x^k = 1 \\ \qquad u^k \cdot y^j - v^k \cdot x^j \leq 0, \quad j = 1, \cdots, m \\ \qquad u^k, v^k \geq 0 \end{cases} \tag{5.23}$$

The optimal weights u^{k^*} and v^{k^*} will give the maximum ratio of $\dfrac{u^{k^*} \cdot y^k}{v^{k^*} \cdot x^k}$ (with $v^{k^*} \cdot x^k = 1$), i.e. the maximum ratio of all weighted outputs versus weighted inputs for port k. By duality of LP, an equivalent *envelopment* form of the DEA model (5.23) can be derived for a given port k, as follows:

$$\begin{cases} \min_{\theta^k, \xi^k} \quad \theta^k \\ s.t. \quad Y \cdot \xi^k - y^k \geq 0 \\ \qquad \theta^k \cdot x^k - X \cdot \xi^k \geq 0 \\ \qquad \xi^k \geq 0 \end{cases} \tag{5.24}$$

where θ^k is a scalar weight on the k-th port and $\theta^k \leq 1$; $\xi^k = (\xi_m^k, \cdots, \xi_m^k)$ is an m-vector weights scaled across all the m ports; $Y = (y_i^j)_{l \times m}$ is an $l \times m$ output (data) matrix observed from all the ports ($i = 1, \cdots, l; j = 1, \cdots, m$); and $X = (x_i^j)_{n \times m}$ is an $m \times n$ input (data) matrix observed from all the ports ($i = 1, \cdots, n; j = 1, \cdots, m$). Given input data X and output data Y, a set of optimal weights for port k, $(\hat{\theta}^k, \hat{\xi}^k)$, can be determined by solving the linear programming problem (5.24), and the minimized value of $\hat{\theta}^k$ by (5.24) is the (technical) efficiency score for the k-th port. If $\hat{\theta}^k = 1$, port k reaches its technical frontier and is regarded as technically efficient. The process can be repeated for all other ports.

Table 5.3 The estimated results of port efficiency parameters

Mean efficiency ($\overline{TE_t}$)	Base model	Model ignoring technical change	Model ignoring unobserved heterogeneity
	Posterior median [5 percentile, 95 percentile]	Posterior median [5 percentile, 95 percentile]	Posterior median [5 percentile, 95 percentile]
1997–1998	0.8072 [0.7009, 0.8293]	0.7015 [0.6221, 0.7832]	0.4138 [0.3376, 0.4880]
1999–2001	0.8393 [0.7406, 0.9153]	0.7816 [0.7202, 0.8675]	0.4267 [0.3652, 0.4825]
2002–2004	0.8423 [0.7415, 0.8986]	0.8602 [0.7867, 0.9267]	0.4344 [0.3676, 0.4925]
Pr($\overline{TE_t} > \overline{TE_{t'}}$)			
Mean efficiency in 1999–2001 is greater than the one in 1997–1998	0.8911	1.0000	0.6634
Mean efficiency in 2002–2004 is greater than the one in 1997–1998	0.8218	1.0000	0.7327
Mean efficiency in 2002–2004 is greater than the one in 1999–2001	0.5446	1.0000	0.6931
Time persistence of efficiencies ($Corr(TE_{it}, TE_{it'})$)	Posterior median [5 percentile, 95 percentile]	Posterior median [5 percentile, 95 percentile]	Posterior median [5 percentile, 95 percentile]
(1997–1998, 1999–2001)	0.6624 [0.2481, 0.8063]	0.7116 [0.5094, 0.8180]	0.8451 [0.7917, 0.8922]
(1997–1998, 2002–2004)	0.2527 [−0.0756, 0.5197]	0.2295 [−0.0878, 0.4995]	0.7264 [0.5941, 0.8020]
(1999–2001, 2002–2004)	0.5511 [0.1947, 0.7566]	0.4452 [0.0995, 0.7017]	0.8459 [0.7877, 0.8874]

Source: Authors.

This chapter has provided a detailed analysis on the effectiveness of port efficiency under the port-focal logistical system. However, as mentioned in Chapter 3, the issue of *environment heterogeneity* cannot be overlooked. In this regard, the influence of institutional factors serves as a highly significant attribute. Moreover, one should not ignore the possibility that the productivity (and strategy) of ports may pose certain externalities on other ports and/or stakeholders along the supply chains.

6
Government Policies and the Role of Institutions

As mentioned in Chapter 4, contemporary developments in shipping forced ports to evolve by undertaking neoliberal management and governance reforms. In theory, such alterations enabled ports to become more integrated into supply chains, while they were also provided an opportunity to transform themselves into maritime logistics hubs with a more hybrid community. However, given the realistic situation within different countries, notably diversity in the institutional system, would a similar solution, when applied to different cases, lead to the same outcome? With the potential pitfalls of increasing governance complexity, was there an international 'best practice' that could fit the new circumstance, or did the choice of appropriate reform approach actually depend on the regional and local circumstances? Moreover, how would institutions affect the port's evolution and its integration into global supply chains, as well as its relations with other components?

Understanding such queries, this chapter discusses government police and the role of institutions in affecting port's evolution, notably its transformation and its integration and relationship with other components of the supply chain. It starts with a brief discussion of changes to the port systems, followed by a theoretical review of the role of institutions on the transformation of economic activities. After then, it provides two cases on how institutions have affected the management and governance structure of two major ports in Asia and Europe, namely Busan (South Korea) and Rotterdam (Netherlands). Information necessary for analysis was collected through 24 semi-structured, in-depth interviews conducted by the authors with relevant personnel in South Korea and the Netherlands, including senior policymakers and industrial practitioners. It provides important theoretical framework in understanding how institutions would affect port governance.

6.1   Evolution of the port system

Researchers have offered a range of perspectives to explain the evolution of the port system (see for instance, Taaffe et al., 1963; Bird, 1971). Some have focused upon the port–city interface (e.g. Hoyle, 1989), while others have developed models based on the competitive nature of port development (e.g. Rimmer, 2007). A widely accepted model of port development was proposed by Hayuth (1981), which divided a port's evolution into five phases. Later, a sixth phase, the so-called 'port regionalization' phase, was proposed by Notteboom and Rodrigue (2005).

According to Hayuth (1981), the first phase of port development took place before containerization and was characterized by scattered ports serving well-defined hinterlands, with inter-port competition being virtually non-existent. In the second phase, ports underwent a process of increasing concentration, with only limited advances in hinterland penetration. This transition was marked by the early stages of containerization at some ports with both strong locational characteristics and an interest in advancing within the established port hierarchy. The third phase was marked by the expansion of containerization and the development of intermodalism. In this phase, feeder ports became increasingly interconnected and dependent on larger ports and hinterland access had expanded further, thus creating overlapping hinterlands and competition with ports located far away from each other. The process of port concentration was replicated inland with increasing concentration and agglomeration at intermediate inland nodes. In the fourth phase, this process was taken to an unprecedented level with the consolidation and widespread adoption of containerization, intermodal transportation and related technological advances. In this phase, integration and speed in cargo flows was advanced by technology, the development of multimodal supply chains and changing shipping networks, thus creating a new hierarchy among ports (like hubs and feeders) and the emergence of load centers. As port-hinterland connections deepened, the emergence of dominant inland ports focusing on major inland market intersections were gradually established, many of which were often connected to ports by high-capacity corridors, both road and rail.

In the fifth phase, load centers grew to a point where de-concentration within the port system started to take place. In this phase, well-positioned feeder ports vying to gain market share would attract ships by offering more favorable terms. This process would not only lead to the movement of cargoes to feeder ports, but also to intermediate hubs – both inland and offshore – to alleviate port congestion. Finally, port regionalization (Notteboom and Rodrigue, 2005) started to develop through deeper integration between gateway ports and inland distribution centers, thus forming regional load center networks with improved efficiency of freight distribution. This phase might reflect a leap in efficiency and in the process of integration of supply chains, where a blurred hinterland became easier to identify through vertical integration between shipping lines, terminals, high-capacity rail corridors and dry ports. This resulted in the creation of a new dimension in port competition where the importance of ports would be measured in relation to their ability to complement and facilitate the establishment of efficient supply chains as they reached further inland.

6.2   A review on the impacts of institutions

Research regarding the impact of institutions on port development has only recently attracted genuine academic interest with the work done by Airriess (2001), Hall (2003), Jacobs and Hall (2007) and Shou et al. (2011), which examined various ports in East Asia and North America. They posited that choices were constrained by institutional conditions and thus led to diversified outcomes and development trajectories. Lee et al. (2008) argued that port evolution in Western, advanced economies was different than what was observed in developing and emerging economies, and there was a risk of implementing the so-called 'western solutions' to developing and emerging economies without first investigating fundamental regional differences. In this regard, transaction cost theory (e.g. Williamson, 1985) suggested that bounded rationality would lead to dominant policymakers advocating the implementation of generic solutions. However, in a cultural political economy analysis, Jessop and Oosterlynck (2008) argued that there was little scope regarding implemented reform as a de-contextualized singularity: when economic forces seek to re-define specific subsets of economic activities, such as subjects, sites, competition and/or objects of regulation and to articulate strategies and visions, they often relied on given authorities to secure certain results. Still, all these were developed within institutionalized boundaries and temporalities in a system of mobilized global capital, that could displace, defer or sustain inherent contradictions.

Indeed, there was a need to balance between economics that naturalized economic categories and soft economic sociology which only focused on the similarities between economic and socio-cultural activities at the expense of the economy's specificity. Historical institutionalism (March and Olsen, 1989; Steinmo et al., 1992; Hall and Taylor, 1998) theorized that institutions posed systematic constraints on individual and collective choices, promoted certain actions and (preferred) outcomes, and pushed non-institutional actors (established based on the original environment, i.e. the grey area in Figure 6.1) towards strategic calculations to optimally fit into the changed environment.[1] Yet, they were rarely the sole cause of outcomes. Ultimately, even the institutional systems depended on what could be done within the economic sphere, notably the pressure along identified directions; the possibility of implementation; the country or region's position within the global economy; and the organization of labor, capital, and the state.

That globalization was still highly segregated, both in definition and impact on economic institutions (Grant, 1997; Stiglitz, 2006), increased governance complexity. Brenner (1998) coined the phrase 'glocal scalar fix,' referring to responses to challenges initiated by global economic development. Reform instruments should be (and were) used differently depending on the differentiation in strategic priorities between authorities locked in diversified institutional frameworks (Henderson et al., 2002). In this regard, neo-institutional theory recognized that path-dependent policy was affected by critical junctures: when events created visions of institutional change and divided events into different periods. The development of institutional change design and evolution was dialectic (Buitelaar et al., 2007). In addition, coercive and persuasive powers enabled the development of events, e.g. governance forms, institutionalized norms and traditions, which countered the change in

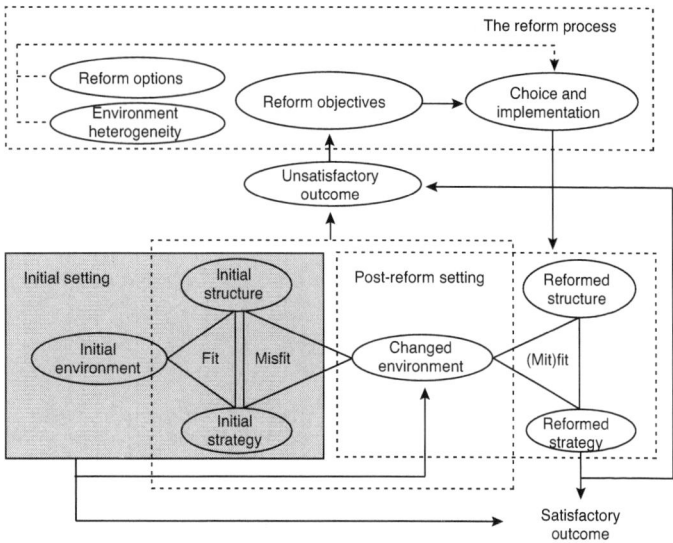

Figure 6.1   The reform process
Source: Authors.

completely de-shaping existing structures. In many cases, reforms within a sector took place incrementally with notable characteristics, reflecting the remnants of the pre-reform setting (Denzau and North, 1994). Unsurprisingly, this might affect the relationship between this particular sector and its surrounding activities, both competing and complementary ones.

The above was valid for organizational change and competitiveness (see, for example, Powell and di Maggio, 1991; d'Aunno et al., 2000; Johnson et al., 2000; Sminia and Van Nistelrooij, 2006), including institutional agents. Various studies focused on changes in the ex-public sectors argued that public organizations were major actors in exercising institutional pressure, but paradoxically also strongly affected by such pressure (Pouder, 1996). When moving service transactions to market environments the prominent role of the regulatory framework, was firmly embedded within distinctively social, legal and economic environments demand institutional studies (Fernandez-Alles and Llamas-Sanchez, 2008).

In the economic sphere, arrangements forming the construct within which a particular sector of the economy operated represented a sub-set of the institutional framework. These arrangements might have had profound effects on the way that the economic sector (or parts of it) in question evolved. Nonetheless, they were only a fraction of the broader institutional framework, with the latter determining the details of these arrangements. Such arrangements existed in cultural and political contexts that defined their form. For instance, institutions might be forced to change by economic and political pressure that was either sector specific or implied by changes in the broader structures of the society.

Apart from the process of change, efficient institutions could add value to assets and promote wealth creation by allowing economic players to invest and specialize. Conversely, inefficient ones would increase transaction costs, e.g. by leading to excessive bureaucracy, corruption, and insecurity among agents, and by reducing the incentives of stakeholders to invest and trade (Coase, 1992). Where aspects of the institutional framework were weak – which was often the case in developing economies – they became vulnerable to manipulation by dominating groups and stakeholders. In this regard, Jessop and Oosterlynk (2008) argued that when economic forces sought to re-define specific subsets of economic activities, such as subjects, sites, competition and/or as objects of regulation and to articulate strategies, projects and visions, they tended to manipulate power to secure certain desired outcomes. Indeed, any governmental failure to provide an appropriate institutional environment might drive economic players to rely on informal, relationship-based, and often less efficient practices (e.g. Martinsons, 2002) which will be further discussed in Chapters 7 and 8.

However, changing institutions might involve dislocating political and economic forces, and thus affect the prevailing balance of power. In this regard, Acemoglu and Robinson (2008) highlighted that by just reforming the institutional structure but not altering the balance of power in society or the basic political equilibrium – the so-called institutional environment – could lead to the replacement of just one instrument by another. According to Hall (2003), institutions changed infrequently but when they did so, it often resulted in dramatic shifts cutting across spatial scales. In the past decades, significant changes in global trade and logistics had forced institutional changes within ports, as operations failed to deliver the levels of flexibility, efficiency and investments required by their users. Such new environments led to increasing uniformity in infrastructure and sectoral arrangements, as port systems were restructured to adopt the new forms of ownership and management in line with generic solutions, notably the *World Bank Port Reform Toolkit* (World Bank, 2007). Essentially, these reforms pointed towards an increasing role for the private sector away from direct public management, with corporatized port authorities being introduced in certain countries and regions. Such convergences, and the success of these reforms, might depend on the geographically confined social environment and institutional arrangements described by Hall (2003) as communities of practice represented a strong countervailing force against change. Studies of various ports around the world (e.g. Airriess, 2001b; Hall, 2003; Jacobs and Hall, 2007) posited that port governance constructs have been defined and restricted by specific institutional conditions, which led to diversified outcomes and development trajectories.

To enhance the understandings on the impact of institutions, this chapter studies the reforms and governance of two ports at the beginning of this century, namely Busan in South Korea and Rotterdam in the Netherlands. Both ports had adopted a generic solution in addressing the global challenges in recent decades (as mentioned earlier), namely the corporatization of respective port authorities which led to the establishment of Port Authority Corporations (PAC). The selection and the examination of these specific ports was in line with the 'corporatization' definition by UNCTAD (1995):

> [Where it] describes the process of which an organization which originally belonged to the public sector is transformed into a corporation with a legal status where the governmental body/ bodies hold(s) the shares of this newly formed corporation.

The two studied cases will be analyzed through descriptive and comparative analysis. The details of the reform process, and the roles of institution factors in affecting the process and outcome will be illustrated.

6.3 Port of Busan

6.3.1 The established institutional system

Following the end of the imperial regime (1910) and the Japanese occupation (1910–1945), the Republic of Korea (hereinafter called 'South Korea') was proclaimed in August 1948 after which it witnessed four decades of authoritarian rule, partly due to perceived threats from the People's Democratic Republic of Korea (hereinafter called 'North Korea') and internal political turmoil. Constitutionally, it was a republic with powers nominally shared among the national president, the legislature, and the judiciary. However, in practice, power was concentrated within the national president (and the executive branch). As noted by Yang (1999), in the post-WWII period South Korea was largely characterized by anti-liberal, anti-parliamentarian and pro- state-developmental policy and institutional system, which frequently clashed with the constitutional pledge of democracy and people's sovereignty. Even after a sustained period of rapid industrialization and urbanization, and following the establishment of the Sixth Republic after the last major constitutional revision in 1988, the executive branch remained powerful; the role of economic freedom and development as a basis of political participation was still virtually lacking (Yang, 1999). Until recently, South Korea fitted well as a developmental state which emphasized state-led macroeconomic planning and the prioritization of economic growth over political reforms. Moreover, the institutional system in South Korea was characterized by 'a strong trunk and weak branches,' in which the national government often played a pivotal role in orchestrating and supervising business (Moon, 1994). Although South Korea adopted an elaborate structure of district and neighborhood (*gu* and *dong*, respectively), local authorities possessed little power, as the functional and financial authorities remained largely within the grasps of the national government. This was perhaps not too surprising due to its state-developmental policies; local authorities were mainly administrative (rather than political) units.

By 2004, there were 51 designated ports in South Korea. Twenty-eight of these (including major ports like Busan, Gwangyang and Incheon) were classified as 'sea' ports, while all others were classified as 'coastal' ports. Until the enactment of the Port Authority Law

(2003), ports were regarded as strategic national public assets. All ports, including ownership, operation, planning and policies were under the direct leadership of the national government through the Ministry of Maritime Affairs and Fisheries (MOMAF). Further, the national government owned all of the port's land and the government was directly responsible for the preparation of the budgets for all the ports. As a consequence, all the ports' extra-, infra- and super-structures were provided by MOMAF and directly financed by public funds. Indeed, the role of ports was very prevalent in the 'national' economic development.

6.3.2   Pressure and motivation for changes

The first major reform of South Korea's ports took place in 1990. Citing management inefficiency, the lack of investment funds (Ryoo and Hur, 2007) and the inadequacy of local inputs and oversight, the national assembly passed an act of parliament to facilitate the establishment of a new governmental department, namely the Korea Container Terminal Authority (KCTA). It acted as the national government's agent in sub-leasing container port facilities (obtained from the national government with zero charge) to private steve-doring firms. Another step of reform took place in 1997 when private firms were given complete freedom in container terminal opera-tions. As a consequence, by 2003, most of the terminal operations had been privatized, with the public sector being limited to public responsibilities. Nevertheless, until then, ports were not authorized to prepare the transport facilities account (a special budget to secure stable construction of port infrastructures) or the general account (the normal budget for daily operations and other necessary minor investments) – both of these accounts (which had direct implica-tions on ports' financial independence) were still maintained by the national government. Moreover, all port facilities were provided by MOMAF, financed by public funds and port incomes.

This situation did not change until 2003 when the national assembly enacted the Port Authority Law – the legal document introducing PACs under MOMAF. Reform details were largely based on the recommendations given by two feasibility studies, the *Government Organization Management Survey Report* (conducted by the national government's task force in 1999) and *A Study on Introducing Port Authority System in Korea* (conducted by the Korean Maritime

Institute (KMI) and the Korean Institute of Public Administration (KIPA), sponsored by MOMAF in 2000). In January 2004, the first such corporation was established in the port of Busan, namely Busan Port Authority (BPA). The motivations for estabishing this corporation were closely linked to the national policy priorities of South Korea. With the rise of production bases in East Asia since the 1990s, an important national policy in South Korea was to grasp this opportunity in order to become a leading nation in Northeast Asia in terms of economic significance. South Korea was unlikely to realize this goal without a strong maritime sector promoting international trade, especially given its relatively small population (49 million by 2006). Land connections between China and South Korea seemed unlikely in the foreseeable future due to the sustained political tangle with North Korea. Thus, the national government of South Korea was keen to make use of its intermediate position in Northeast Asia to develop the country into a logistical hub, and MOMAF was responsible for transforming the country's major ports in order to achieve this ambitious objective. As indicated in MOMAF's briefings, port policy direction was developed in close accordance with national policies, e.g. concentration on container cargoes, development of port backup areas to attract trans-shipment traffic, cooperation with China and Japan, and so on (MOMAF, 2004).

Simultaneously, intensified inter-port competition in Northeast Asia raised the fear that the port of Busan (and other South Korean ports within the region) would lose their competitive position to Chinese ports. It was recognized that terminal privatization alone, which only emphasized efficiency enhancement, seemed inadequate to tackle the challenges presented by intensified port competition. In fact, from 1996 to 2003, the growth of the port of Busan was persistently lower than its Chinese counterparts (Containerisation International, 2006). As a consequence, the South Korean national government perceived that its traditional port governance system, characterized by a high level of bureaucracy, was inadequate to respond to the rapidly changing demands placed on port. Further, the relative success of port reforms by its neighbors (like Chinese ports, see Shou et al., 2011) convinced the South Korean government that further reforms must take place so as to avoid losing out. The national government responded to these national and global trends with the formation of PACs and BPA.

Furthermore, since the 1990s, port reform was part of the national decentralization processes in the South Korean society as a whole, partly due to increasing requests from regional and local authorities to enhance their roles in the decision-making process of major issues, including port development. Although 'landlord' ports (cf. World Bank, 2007) had largely been established, port governance was still dominated by the national government. Despite port-initiated negative externalities on its surrounding regions like traffic congestion and pollution became increasingly visible, the sharing of power with local authorities (notably municipal governments) hardly existed.

6.3.3   Structure and function of BPA

The reformed structure of BPA can be found in Figure 6.2. As a newly established corporation by the national government, the objectives of the BPA were double-folded. On the one hand, it had to ensure that the port of Busan would contribute significantly to national economic growth, while on the other hand it should develop the port into a competitive logistics hub with efficient and optimal services, and expanding infrastructure (BPA, 2006). The top hierarchy consisted of a supervisory Non-Executive Board, the Port Committee, with

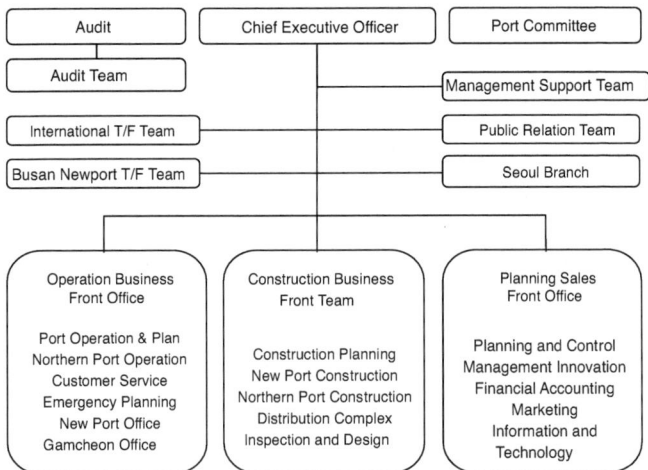

Figure 6.2   The structure of Busan Port Authority (BPA) at its inauguration
Source: BPA (2006b).

members representing various sectors within the maritime community. In 2005, the Port Committee had 11 members appointed by MOMAF (with MOMAF and the Busan municipal government nominating six and five members, respectively). Executive power was concentrated in the Chief Executive Officer (CEO) nominated by the Port Committee, with advice from MOMAF and the mayor of the city of Busan, and duly appointed by the national president for three years (with a possible re-appointment). Through its three main branches (Planning Sales, Operation and Construction Businesses), BPA was responsible for implementing most of Busan port's commercial and developmental functions, while MOMAF, through various Regional Maritime Affairs and Fisheries Offices (RMAFO), was to maintain public functions, notably safety and security-related issues.

While initially composed of successive employees from MOMAF, RMAFO and KCTA, in order to dilute the port authority's 'public' image and to strengthen marketing, public relation and new projects, BPA recruited numerous employees with private sector backgrounds, expanding to 17 teams and 146 permanent staff members by 2006 (compared to the original 11 and 77, respectively) (BPA, 2006). Financially, BPA adopted a self-supporting budgeting system that emphasized balancing income and expenditure (with separate balance sheet records), a departure from the traditional general budgeting system where the port's finances were inscribed into the country's national budget. Being granted financial autonomy encouraged BPA to be more responsive to business environments, while simultaneously reducing public influence on day-to-day operations. Financial autonomy also implied that BPA should assume substantial responsibilities for port facility investments. From 2004 to 2006, BPA's net income grew from nine to more than 16 million US dollars (BPA, 2006).

Apart from separating public-private responsibilities, the aforementioned reform also readdressed the national-dominated system (Yeo and Cho, 2007). The establishment of BPA had witnessed increased participation from Busan's municipal government on port affairs, mainly in terms of supporting BPA and the port of Busan's development through financial incentives. For example, in 2004, BPA was granted profit tax exemptions from the municipal government for three years. Also, as mentioned earlier, for the first time, Busan's municipal government, through the Port Committee, was realistically involved in the nomination of the port's CEO. However, the limitations of municipal government in port affairs could be easily recognized. Despite its call for a municipal port authority, which was supported by the *Government Organization Management Survey Report*, BPA was still under the direct control of the national government, as illustrated by the government's refusal to allow BPA to issue shares. Moreover, while allowing the municipal government to participate in the appointment of Port Committee members, by the time BPA was inaugurated, the CEO (with full executive authority) was still directly appointed by the national president. Figure 6.3 illustrates the power relationship within the port of Busan since the BPA's inauguration in the mid-2000s.

The above illustrated that the South Korean national government desired to decentralize some of its power to the local authority, while

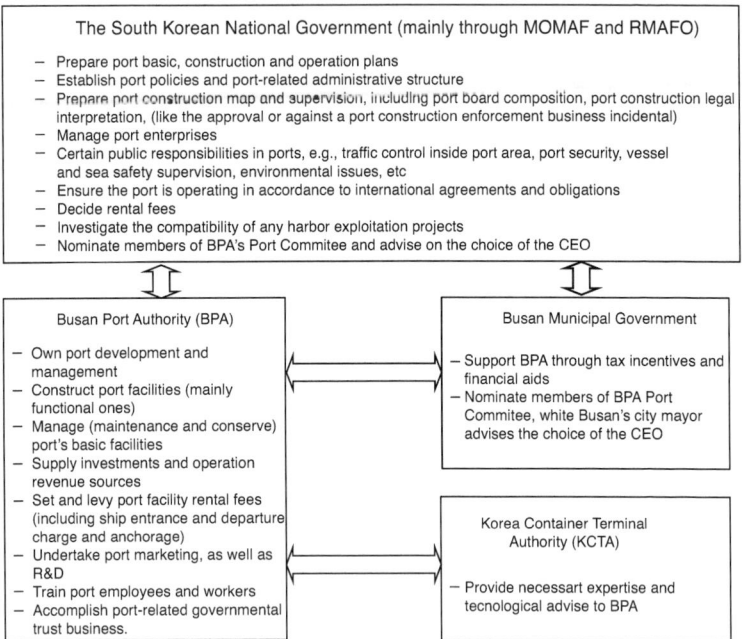

Figure 6.3   The power relationship within the port of Busan since Busan Port Authority's (BPA) inauguration

Source: Authors.

it simultaneously attempted to further separate commercial and public functions. In this regard, Busan port's reform highlighted the change of policy direction in port development, transformed from the national government-led governance system to a more diverse governance system, of which some kinds of distribution of responsibilities and power-sharing existed between different port stakeholders. Nevertheless, while it had done more than enough in terms of commercialization, the new system was still very much in line with the traditional institutional system, as characterized by 'a strong trunk and weak branches'. The national government still played a pivotal role in orchestrating and supervising business, and virtually all the top appointments of BPA reflected strong political, rather than commercial, characteristics. Most key BPA decisions remained firmly in the hands of the national government, while the core port projects still largely depended on the national government's financial muscle. Instead of being purely port-oriented, the corporatization of BPA was a means to fulfill the objectives of the national political agenda in developing South Korea as the regional logistical hub in Northeast Asia.

6.4   Port of Rotterdam

6.4.1   The established institutional system

The Netherlands was among the founding member-states of the European Union (EU) with a strong pro-market tradition insofar as the economy was concerned. The advocacy of the so-called 'four freedoms,' i.e. the free movement of services, labor, capital, and the freedom of establishment within Europe, and the strong autonomy of local government, quintessentially challenged the notion of an interventionist state which centrally detailed a plan regarding the development for each sector of the economy. Indeed, local authorities in the Netherlands conformed to the basic logic laid down in 1848–1851 (Constitution 1848; Municipal Act 1851), the basic logic been commonly described as the logic of the decentralized unitary state. The unitary element was not so much based on hierarchical steering but rather on a mutual adjustment between the three active levels of the state: national government, municipality (provincial) government and local government (Toonen, 1990). Local and municipality governments complied with the principle of bounded autonomy, having an open, general competence. The stability of the system remained (i.e. the internal structures), though the rapid adoption of the New Public Management (NPM) in the 1980s, implied that throughout the country municipalities had experimented with self-management, contract management and related forms of business like practices (Hendriks and Tops, 1999). NPM also introduced the idea of the municipality as a 'holding' allowing the flourishing of a number of 'product divisions' under the municipality (Camps, 1996). Also, it reformed the bureaucratic culture, and emphasized entrepreneurship, i.e. indicators and benchmarking infiltrated the municipalities, perusing the replacement of what was considered as a highly procedural and risk-reducing culture (Pauka and Zunderdorp, 1990).

Hence, the Dutch port system was characterized by a decentralized system with municipalities taking care of respective ports. Although there was national legislation on public-related issues (like safety, security, customs, pilotage and environmental protection), an overall national port management system did not exist. Traditionally, the task of the national government, represented by the Ministry of Transport, Public Works and Water Management (MVW) and the National Port Council (NPC), was to respond to the initiatives of port authorities and industries, and to provide extra-port infrastructure, notably water channels. Until 2003, all major decisions on port affairs in Rotterdam, including port bye-laws and port dues, were made by *Gemeentelijk Havenbedrijf Rotterdam*, or Rotterdam Municipal Port Management (RMPM), while similar systems were also adopted in other ports, e.g. Amsterdam, Moerdijk, Vlissingen, and so on (Dekker, 2005). It followed the Hanseatic tradition of the 'landlord' port authority and of the powerful presence of the local or municipal management.[2]

The structure of RMPM by 2003 can be found in Figure 6.4. It was noted that the Rotterdam Port Authority (RPA) had the same function as the corporatized PoR's Harbor Master (to be discussed later). While there was a division on commercial affairs, as a department affiliated to Rotterdam's municipal government, RMPM undertook its role as a landlord rather than commercial functions. On the one hand, although RMPM had financial autonomy, money flows were not clearly distinguished between public and private functions. The national government provided extra-port infrastructures through public funds as long as it considered such infrastructures as contributing to national economic growth (RMPM, 2002). On the other hand, port infrastructure was financed by RMPM through determined rent (long term lease), quay charges and port dues. Superstructures were privately owned, and financed, by terminal operators through operating revenues. In some cases, they were responsible for specific adaptations of the terminal area including, for example, the construction of pavement and crane tracks.

6.4.2   Pressure and motivation for changes

In January 2004, RMPM's Commercial Affairs Department and RPA were detached from RMPM to form a PAC, namely Port of Rotterdam Authority N.V. (*Havenbedrijf Rotterdam N.V.*) (PoR); it was incorporated for 800 million Euros (about one billion US dollars). The management, accountability, supervision and financial operations of PoR would be carried out on the basis of conventional company law, articles of association and internal regulations of the country (no port-specific laws were introduced). The major focuses of PoR were commercial and financial affairs, including investments of new development projects, re-development of old port areas and getting new customers.

Two main reasons prompted the reform in Rotterdam.[3] First, it was due to changes in the global economy and increasing port

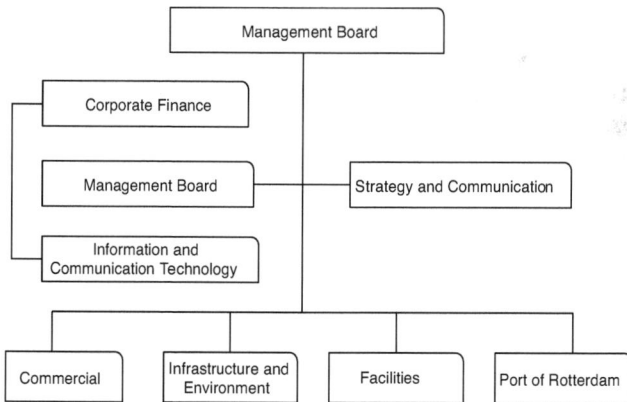

Figure 6.4   The structure of Rotterdam Municipal Port Management (RMPM) by 2003

Source: RMPM, 2002.

competition. Since the mid-1990s, it was recognized that Rotterdam port persistently grew slower than its major competitors, especially in container traffic. This situation became worse in the late 1990s, when Rotterdam experienced negative growth for several years (Containerisation International, 2006). The implications were clear: with intensified inter-port competition, the 'landlord' port of Rotterdam could not sustain its competitiveness through the privatization of terminal operations alone. Indeed, a major stumbling block of the landlord port was its ignorance of the fact that terminal operators could/would not carry out the commercial functions related to the port as a whole, e.g. enhancing port's image and reputation, port investment projects, and so on. Until 2003, the marketing of the port was still undertaken by RMPM.

Secondly, the changing relation between Rotterdam port and government, especially the national government, due to the progress of European integration, had become significant. It became increasingly difficult for the port to make its own decisions due to the various effects of European integration, effects derived by both the presence of a single European market and the increase in the EU's involvement in port issues (Chlomoudis and Pallis, 2002; Pallis, 2007). Given the port's importance to the Dutch economy,[4] the national government was concerned about the lack of the port of Rotterdam's involvement in any EU policies, which could threaten its competitive position and possibly hinder national interests (MVW, 2005). For instance, ports and supply chains could operate efficiently only if they had good hinterland connections and transport infrastructure projects – like the *Betuweroute*, a (at the time) proposed freight-only rail connection linking the port of Rotterdam to Germany and other inland European destinations (PoR, 2007) – which not only required financing but also the cooperation of other national authorities and the EU. Further such projects would be potentially difficult to finance without the support of the national government. Without the national government's participation, Rotterdam port would find it difficult to fulfill the objectives of its ambitious long-term port plan (*Port Vision 2020*), especially those aspects involving its international competitiveness through improving its accessibility. On the other hand, greater port autonomy would help port entities to implement relevant supranational legislation and conform to new (de)regulatory requirements. The reform served as the initial step to enable PoR to develop a long-term international strategy in port management and governance around the world (Dooms et al., 2013).[5]

6.4.3 Structure and function of PoR

As a public corporation, PoR has two key objectives: (i) to promote the effective, safe and efficient handling of shipping and to arrange for nautical and maritime order and safety; and (ii) to ensure that the port of Rotterdam could be managed and operated in a professional manner – not necessarily the cheapest, but the best in reliability, service quality and competitiveness (PoR, 2004). The new management structure was designed in a way that PoR would be sensitive to cost, opportunities, customer satisfaction and social responsibilities.

The structure of PoR can be found in Figure 6.5. It consisted of a non-executive board with five to seven members (depending on circumstances). Its major responsibility was supervisory and to ensure that PoR could operate effectively as a business while maintaining public accountability. Its composition must reflect its dual nature (PoR, 2004). PoR shareholders elected members during the general meetings, including the board's chairman and the commissioner, and no one could serve within a particular capacity for more than a dozen years. By doing so, the executive board had adopted a collective decision-making system consisting of six members: the CEO, Chief Commercial Officer (CCO), Chief Financial Officer (CFO), Chief Operational Officer (COO) (CCO, CFO and COO led the Directors of Commercial Affairs, Finance & ICT and Port Infrastructure & Maritime Affairs Directorates, respectively), the Harbor Master and the Managing Director of *Maasvlakte 2*. The CEO was responsible to make and execute all major decisions in consultations with other members of the board. Members of the executive board were nominated by the non-executive board and appointed by the commissioner, with no specific term limitations.

In terms of human resources, a significant restructuring took place to adopt smaller but more specialized task-oriented work units. New collective agreements had been adopted with the labor union and new private employee contracts had been agreed upon with the 1,200 RMPM staff. From 2002 to 2004, RMPM's staff population

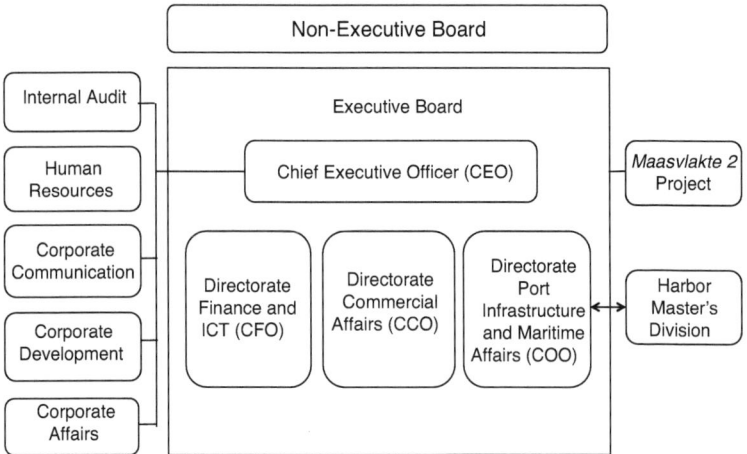

Figure 6.5   The structure of the Port of Rotterdam Authority N.V. (PoR) at its inauguration

Source: PoR, 2006.

increased consistently, while the trend has been reversed dramatically since then (as PoR) (Annual Reports of RMPM (2001–2003) and PoR (2004–2005)). On the contrary, the number of divisions persistently increased since corporatization. In 2003, RMPM only had four divisions, compared to 25 divisions under PoR by 2006. By adopting more task-oriented work units, it was hoped that outcomes could be more effectively evaluated, thus enabling performance-based remuneration policies to take place. Furthermore, PoR needed to prepare annual business plans to address how mid-term business plans (like *Port Vision 2020*) could be carried out, with a view to complement how long term objectives of Rotterdam port could be fulfilled.

As mentioned before, the national government traditionally played a marginal role in port affairs, despite the fact that the port of Rotterdam's operation alone consistently made a substantial contribution to the country's gross national product (GNP) (PoR, 2006). With the establishment of PoR, however, the situation changed. Despite objections from the Rotterdam municipal government, the first initiative took place in 2006 where the national government bought 25% of PoR's shares. Later, to act as a new source in financing *Maasvlakte 2*, in 2008, the national government planned to invest further and increased its PoR shareholding to about 30% (PoR website, 2013), implying that the national government would have a more explicit role in port of Rotterdam's matters, including the appointments of board members and project investments. More importantly, the national government started to act as the liaising agent between PoR and the EU, especially on issues about the role of the port of Rotterdam in various EU transport policies, like the Trans-European Transport Network (TEN-T). The emergence of supranational, i.e. the EU, policies contributes to a more active role of the national administration.

Despite such developments, the presence of the municipal government in PoR remained strong; RMPM remained the largest shareholder (about 70% by 2013). Also, as a compromise pre-condition for the establishment of PoR, the latter must take up all the port-related public obligations performed by RMPM before PoR was created, e.g. marine safety, infrastructure maintenance, environmental management, and port security. This was accomplished by transforming RPA into a new department within PoR, namely Harbor Master (directly appointed by Rotterdam's municipal government). Finally, PoR must finance all facilities that RMPM financed before, notably port infrastructure, while the ownership of port's land firmly remained within Rotterdam municipal government, with PoR paying billions of Euros worth of dividends to the municipal government annually as lease (while PoR leases port's land to individual terminals). Other financial arrangements included: (i) nearly one billion Euro loans with Rotterdam's municipality must be re-financed within a decade. PoR pays off 10% of its debt each year by re-financing; and (ii) an annual sufferance tax of 12 billion Euros (about 16 billion US dollars) levied on PoR for facilities, e.g. chemical pipes, cables and wires. Figure 6.6 illustrates the power relationship within the port of Rotterdam since PoR's inauguration in the mid-2000s.

The establishment of a PAC in the port of Rotterdam illustrated a decentralized system, of which the dominance of a particular player in the governance system was not as visible as in the port of Busan. PoR was largely exempted from political interference and could handle its own affairs, while governments, both national and municipal, mainly played supporting roles. PoR was largely structured like any other normal businesses and even with specific restrictions to minimize political intervention in its strategic decisions. While there was some power-sharing and functional distribution between different governmental levels, PoR did not make any serious attempts to establish formal liaisons with the EU – it was largely left to the national government (unlike some countries like Greece, see Ng and Pallis, 2010). The relative freedom of the port of Rotterdam from political interference was exemplified by the fact that its port plan at that time – *Port Vision 2020* – was evidently more closely addressing the future of itself (and city of Rotterdam) rather than the national political agenda. Indeed, based on anecdotal information, PoR was much more interested in the port's international competitive position, rather than how it could contribute to achieve the objectives of the national political agenda. Even when participating in development projects (like *Maasvlakte 2*), unlike South Korea,

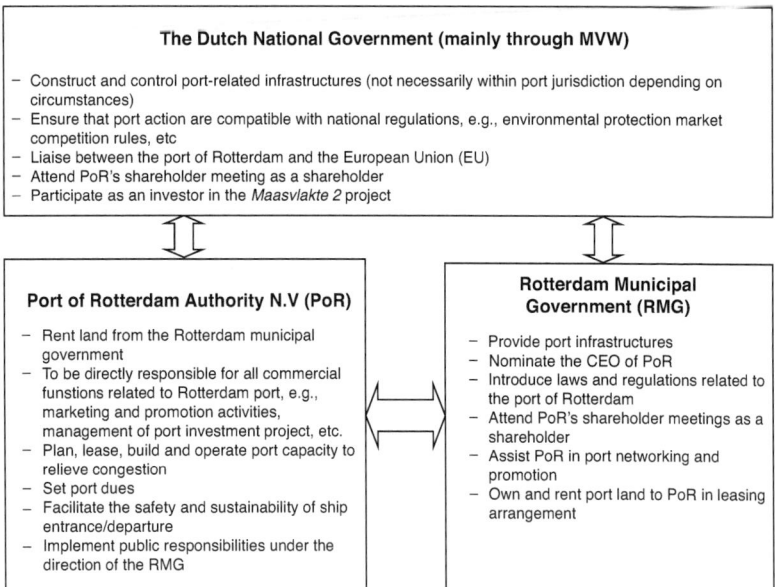

The Dutch National Government (mainly through MVW)

- Construct and control port-related infrastructures (not necessarily within port jurisdiction depending on circumstances)
- Ensure that port action are compatible with national regulations, e.g., environmental protection market competition rules, etc
- Liaise between the port of Rotterdam and the European Union (EU)
- Attend PoR's shareholder meeting as a shareholder
- Participate as an investor in the *Maasvlakte 2* project

Port of Rotterdam Authority N.V (PoR)

- Rent land from the Rotterdam municipal government
- To be directly responsible for all commercial funstions related to Rotterdam port, e.g., marketing and promotion activities, management of port investment project, etc.
- Plan, lease, build and operate port capacity to relieve congestion
- Set port dues
- Facilitate the safety and sustainability of ship entrance/departure
- Implement public responsibilities under the direction of the RMG

Rotterdam Municipal Government (RMG)

- Provide port infrastructures
- Nominate the CEO of PoR
- Introduce laws and regulations related to the port of Rotterdam
- Attend PoR's shareholder meetings as a shareholder
- Assist PoR in port networking and promotion
- Own and rent port land to PoR in leasing arrangement

Figure 6.6 The power relationship within the port of Rotterdam since the inauguration of the Port of Rotterdam Authority N.V. (PoR)

Source: Authors.

the Dutch public sectors employed a rather 'business-as-normal' approach *via* financial investments and share purchasing. All these factors suggest that PoR's projects were carried out closely in accordance to the Dutch institutional system which emphasized liberalism and minimal intervention in business, including ports.

6.5 A comparative analysis and the role of institutions

It is possible to identify a number of commonalities between the reform nature of the studied cases. First, despite their differences within the institutional system, both case studies highlight the decision to include the previously peripheral players into the management system, as a means to tackle new global challenges. As a result, the reformed structure of both cases had shared similar characteristics, i.e. the beginning of participation by the initially peripheral players. Some compromises or power sharing was witnessed, and both of them entered a new phase with a more complex governance system, with more players, being established.

Further commonalities could also be found, with financial autonomy being the major center of attention. Recognizing the perils in jeopardizing PAC's autonomy and real changes, both BPA and PoR were granted (limited) financial autonomy, and allowed to prepare independent budgets. Finally, both of them made some deliberate attempts to dilute the 'public' image as a port authority, as indicated in their policies regarding new staff employment and contract renewals which emphasized removing bureaucracy, a more task-oriented focus, as well as being more commercially minded. Nevertheless, when the detailed content was magnified, it was found that the reformed structures between them were largely diversified and, interestingly, in accordance to respective established institutional systems, as illustrated below.

6.5.1 Corporate nature

A comparison of the PACs, including the supporting legal documents and shareholding structure, can be found in Table 6.1.

As illustrated in Table 6.1, despite being a PAC, the South Korean government clearly attempted to preserve the nature of its institutional system. Rather than being governed by company and business laws, the new structure was legally supported by special laws

Table 6.1 The legal and shareholding structures of Busan Port Authority (BPA) and the Port of Rotterdam Authority N.V. (PoR)

Category	Busan	Rotterdam
Name	BPA	PoR
Legal document	Port authority law	No port-dedicated law existed (governed by conventional private company law)
Shareholding	No	Yes

Source: Authors.

enacted by the national parliament. Moreover, while the establishment of BPA was initiated by the national government, the government repeatedly refused to issue shares. On the contrary, in the port of Rotterdam, the structure of PoR reflected a much more business-oriented nature. For example, PoR was a shareholding firm, as a result, the management, accountability, supervision and financial operations of PoR would be carried out on the basis of conventional Dutch company laws, articles of association and internal regulations (no port-specific laws were introduced). This business nature could be found in its non-executive board (NEB), of which no politicians or members from interest groups were allowed to serve as NEB members. This was clearly lacking in the port of Busan. In BPA, the composition of the supervisory institution clearly reflected diversified political interests, notably the interest groups, labor union and dockworkers (Table 6.2).

As shown in Tables 6.2 and 6.3, the major functions of Busan's supervisory institution, the Port Committee, was very much restricted to offering advice and to nominating the CEO. It had no real power in auditing the executive branch, and was not authorized to appoint or remove the CEO from office by the time that BPA was inaugurated. On the contrary, the non-executive board of PoR could play a much more check-and-balance role. It was not only authorized to appoint the CEO, but also able to suspend him/her from duty, if deemed necessary.[6] Indeed, even after the establishment of BPA, the executive power was still highly concentrated. Within the organization, the only authorized executive member was the CEO who was responsible to decide upon and carry out all executive functions. By the time when BPA was

Table 6.2 The supervisory institution of Busan Port Authority (BPA) and the Port of Rotterdam Authority N.V. (PoR)

Category	BPA	PoR
Name	Port Committee	Non-Executive Board
Composition of supervisory board	6 interest groups; 4 academia; 1 labor union	Depends on circumstance, but no politicians/members from interest groups
Functions of supervisory board	Advise the choice of the Chief Executive Officer (CEO) to MOMAF and the national president	Nominates and appoints Executive Board members, including CEO, as well as their removal, if deemed necessary

Source: Authors.

Table 6.3 The key executive appointments within the Busan Port Authority (BPA) and the Port of Rotterdam Authority N.V. (PoR)

Category	BPA	PoR
Name	(N.A.)	Executive Board
Appointment of CEO	National President	Non-Executive Board
Appointment of other executive members	(N.A.)	Non-Executive Board

Source: Authors.

inaugurated, it was also only accountable to the national president. With such a setting, the national government did not even seem enthusiastic to give complete freedom to BPA in its development projects where some commercial functions within port should remain within a public framework to ensure that strategies and actions would be in accordance with the 'national' welfare, as exemplified by the continuing existence of various departments within the restructured RMAFO (Port Logistics Division, Port Construction Division and Gamcheon Branch Office) with many responsibilities largely duplicating BPA's functional branches of Business Operations and Construction Business. On the contrary, the CEO of PoR was less powerful due to the existence of an executive board within the senior hierarchy. Moreover, it was clearly stated that the CEO should not act alone, but needed to make and execute all decisions in consultation with other members of the executive board. Except the Harbor Master, all other departments within PoR were business-oriented, both in terms of structure and personnel.

6.5.2 Power sharing between different institutional hierarchies

The power and responsibilities of national governments within their respective PACs and ports in the mid-2000s can be found in Table 6.4.

As mentioned, in accordance with the Dutch institutional system, the national government played a peripheral role in port affairs. Despite acquiring 30% of PoR's shares, it was still nearly uninvolved in the operation and development of the Rotterdam port, apart from ensuring that port operations and development was in compliance with national (and the EU) laws and regulations, as well as acting as the liaison agent between the port and the EU (as this was out of the

Table 6.4 The power and responsibilities of the South Korean and Dutch national governments within Busan Port Authority (BPA) and the Port of Rotterdam Authority N.V. (PoR), respectively, in the mid-2000s

Category	Busan	Rotterdam
Shareholding within PAC	–	30%
Ownership of port's land	Yes	No
Infrastructure construction	Fully involved	Not involved
Introduction and enactment of Port-related laws and regulations	Fully involved	Only ensure that port operation and development is in compliance with national (and the European Union, or EU) regulations
Assistance in port networking and marketing	Partially involved	Not involved
Involvement in port development projects	Significant	Limited
Others	–	Liaison between the port and the EU

Source: Authors.

port's jurisdiction). This was in stark contrast to the South Korean case. The South Korean national authority was directly involved in nearly every aspect of port development, including ownership of the port's land, infrastructure construction, and the introduction and enactment of port-related laws and regulations. Furthermore, it was actively involved in Busan's networking and marketing, and played a direct role in the development of Busan's development projects, notably the Busan Newport (BNP). Given such diversity, it was perhaps not surprising to find that significant differences could also be identified in the power and responsibilities of municipal governments, as illustrated in Table 6.5.

Despite the call for a municipal-level port authority, which was supported by the *Government Organization Management Survey Report*, the limitations of municipal government in port affairs could still be easily recognized and the Busan municipal government was unable to play a role of any significance in BPA. Apart from providing tax incentives to BPA (like tax exemption in 2004–2007), the

Table 6.5 The power and responsibilities of the Busan and Rotterdam municipal governments within the Busan Port Authority (BPA) and the Port of Rotterdam Authority N.V. (PoR), respectively, in the mid-2000s

Category	Busan	Rotterdam
Shareholding within PAC	(N.A.)	70%
Ownership of Port's Land	No	Yes
Construction of Infrastructure	Not involved	Fully involved
Introduction and Enactment of Port-related Laws & Regulations	Not involved	Fully involved
Assistance in port networking and marketing	No/trivial involvement	Active involvement
Involvement in Port Development Projects	Limited	Limited
Others	Tax incentives to BPA; Nominates Port Committee members; Busan Mayor gives advice to BPA's CEO appointment	–

Source: Authors.

nominations of Port Committee members and the choice of CEO to the national government were already the main 'authorities' that the municipal authorities could implement. Moreover, it did not own any port's land, nor had it been involved in any real port-related development, like infrastructure construction, port promotion or marketing. Despite reforms, the national government-dominated system (Yeo and Cho, 2007) seemed to sustain in Busan by the time that BPA was inaugurated. On the contrary, the Rotterdam municipal government retained the ownership of port's land, while it also played pivotal roles in port infrastructure construction, as well as its networking and marketing. As the majority shareholder the Rotterdam municipal government was also the key player in deciding the nomination and appointments of key personnel within PoR as discussed earlier.

6.5.3 The role of institutional agents in port development projects

The differences between the studied cases can be further illustrated by their government's approach in respective port development projects (Table 6.6).

Table 6.6 The roles of the South Korean and Dutch national governments in the port development projects planned and managed by the Busan Port Authority (BPA) and the Port of Rotterdam Authority N.V. (PoR), respectively, in the mid-2000s

Category	Busan	Rotterdam
Name of project	Busan Newport	Maasvlakte 2
Motivation for government funding	National priority project	As part of the funding of public goods, e.g. sea defense
Method of funding	Direct investments	Purchasing PoR's shares
Funded amount	€ 3.3 bn (45%)	€ 0.5 bn (19%)
Items eligible for government funding	No limitations	Public infrastructure only
Environmental and social issues	Limited significance	Great significance

Source: Authors.

As illustrated in Table 6.6, significant differences can be found on how the national governments supported their respective ports. The South Korean government financed nearly half of the budgeted amount of Busan's port development projects through direct investments, taken out of the national budget, while for *Maasvlakte 2*, the Dutch national government mainly treated it as a conventional business, and invested only through the purchase of PoR's shares (and a substantially smaller percentage, only 19% of the budgeted amount). Moreover, it paid much attention to ensure that the port of Rotterdam would not gain significant competitive advantages due to public funding. Indeed, it even stated that public investments could not be commercially exploited or able to generate any revenues for PoR (*Maasvlakte 2*, 2013).

The Dutch's port-oriented approach was different from BNP. The objective of BNP, as an ambitious project launched by the South Korean government itself, was to meet the forecasted rapid growth of container traffic. In this case, limitations on construction items and commercial advantages, like what the Dutch did, were virtually non-existent and the national government was directly involved in the development of BNP (Ryoo and Hur, 2007). For example, in BNP's Phase 2, four ship berths were financed by public funds so as to enable the development of 13 ship berths (BPA, 2007). The following quote from PNC's official website has illustrated this point extremely well:

> Facilities supported by government [including] breakwater, disposal area, road and railroad connected to [BNP] terminal... to ensure that the logistical functions of Port of Busan will be second to none in the region, BPA will construct a hinterland to the rear of the new port of Busan (1.1 million sq. m) in cooperation with the central government. (Pusan Newport Co., 2007)

In Busan, BNP was actually a national priority project which remained high in the national political agenda, with the objective of transforming South Korea to become the regional logistics center in northeast Asia. It was the national government, not BPA, which acted as the 'engine room' for this project. As indicated in MOMAF's briefings, port policy direction must be developed closely in accordance with the national political agenda, being driven as one of the priority national policy projects (Ng and Pallis, 2010). Indeed, even Pusan Newport Co. (the subsidiary company established under BPA to manage the construction of BNP) admitted that the purpose of BNP was to enhance the Korean national competitive power through the enlargement of port facilities (Pusan Newport Co., 2007). Given the substantial amount being pumped into BNP by the national government, it was conceived of as more than just a pure port development project. Thus, it was sensible to argue that BNP is a national tool in achieving national objectives, with BPA very much an agent to help the national authority in achieving such ambitions. On the contrary, *Maasvlatke 2* was clearly a port-oriented project in pushing the port of Rotterdam's ambition to become a European, and even global, maritime logistical hub. Indeed, such a difference reflected the more 'international,' rather than national, ambition of PoR as mentioned earlier,[7] and it was recognized that the connection between the port of Rotterdam's development and national objectives was not explicit at all. Quoting *Maasvlakte 2's* website:

> Now that [the Dutch] parliament has taken the 'go' decision [for Maasvlakte 2] this week, the Maasvlakte 2 project is in fact no longer on the national political agenda. (Archived news in 2006, found in *Maasvlakte 2* website)

Analysis of the approach dealing with the port development project was pivotal in distinguishing how the institutional system impacted port governance. To a large extent, *Maasvlakte 2* remained port-oriented, and the role of government was largely supportive rather than directive. On the contrary, BNP was very much detached from BPA and Busan port, its core aim was to fulfill national, rather than port or local, ambitions.

6.6   Discussion and conclusion

The studied cases confirmed that governmental institutions constrained actions, though they were not necessarily the sole cause of outcomes. As illustrated in both studied cases, exogenous factors like technological and economic progress led to the corporatization of the respective (initially public) port authorities. Implementing adjustments within different institutional frameworks, relevant policymakers in both ports had actually moved forward towards a similar direction in management and governance structure, and new systems aimed at financial autonomy, diluting the 'public' image and the inclusion of previously peripheral stakeholders in forming a network of self-governing actors which all participated in port governance, and the future direction of port development. Indeed, the outcome of the idea of establishing PACs depended heavily on the institutional systems of respective public administration and their capacities to implement such changes.

Thus, this chapter has offered an interesting comparative study to identify how different countries, with different institutional systems, might implement a generic approach in solving a globally proximate problem. Although on the surface both Busan and Rotterdam moved towards a converging direction in terms of governance, the reality was that both of them failed to move away from their established institutional systems. South Korea remained state-developmental, while the Netherlands continued to adopt its entrepreneurship approach. In this regard, Rotterdam's entrepreneurial approach could even be dated back to the mid-twentieth century when containerization started to take off, where the limited role of the municipal government in port superstructures forced the city's firms to play highly active roles in port facility investments, as well as the early participation of large stevedoring companies in the construction of container-dedicated terminals in the port of Rotterdam (Vanfraechem, 2012). There was little doubt that both Busan and Rotterdam had been 'locked-in' (Pierson, 1993) by their own restraining institutional systems.

This chapter offered a significant contribution to the understanding of port evolution and its potential relationship with other components of the supply chains, especially in view of an increasingly complex port system. It strongly indicated that institutional systems would be a key variable in defining the transformation, evolution and development of contemporary ports and multimodal supply chains. More importantly, the diversification of the internal structures of different ports hinted that they might pose diversified impacts on the integration process of ports into logistics and supply chains around the world. This would be especially prominent in developing economies where the influences of institutions were usually strong. Understanding such, through investigating two of the world's most important developing economies, namely India and Brazil, the evolution and relationship between ports and other components of logistics and supply chains, notably dry ports or inland terminals, will be further discussed and analyzed in Chapters 7 and 8.

7
Case Study – India

This chapter examines one developing economy, and investigates the development of dry ports in India over the past two decades, focusing upon how institutions and government policies acted as the *environment heterogeneity* in affecting the geographical settings of such facilities. The chapter will discuss the dualistic policies undertaken by the Indian government in its dry port sector, and how such policies have affected the spatial characteristics of dry ports in India. With more than 200 dry ports around the country, are all of them fully operational? Is there any wastage of resources? Why are some dry ports more successful than others (at least from an operational perspective)? This chapter will provide insight on the above queries. Towards the end, three case studies located in Southern, Central and Northern India will be examined, so as to illustrate the impact of institutional factors in the spatial settings of India's dry ports. It should be noted that much of the information and analysis in this chapter was based on 26 semi-structured, in-depth interviews conducted by the authors with relevant personnel during a field trip undertaken in 2009–2010 (hereinafter called 'interviewees' while the unpublished information provided by them were called 'anecdotal information').

7.1 A brief review of the development of Indian dry ports

India, one of the fastest growing emerging economies in Asia, is facing a major challenge as it continues to pursue its policies of stimulating economic growth and development. These challenges stem from an underdeveloped and inefficient land transport infra-structure that has not kept pace with rising demand and has thus constrained India's economic progress. India's road network, which carries almost 90 percent of its passenger traffic and 65 percent of its freight, is congested and of poor quality. Although the density of improved roads (0.66 km per km^2 of land) is almost identical to that of the US (0.65) and much greater than in China (0.16), most of India's roads are narrow and congested, with poor service quality (World Bank, 2010b). Its railway network, carrying some 17 million passengers and two million tons of cargo per day, is inflicted with capacity constraints and operational inefficiencies. All these factors, together with the prioritizing of passenger traffic over freight traffic within its rail system, have raised India's freight tariffs to among the highest in the world (Bajas, 2010). In addition, the required invest-ment and modernization of its railways has been hampered by the need for development policies to be consistent with the political real-ities of India's consultative democracy, unlike the situation under a more authoritative and directed developmental approach, such as China's (Prime, 2009; Lin, 2010). In this context, the reinforce-ment of India's container ports, and in particular its dry port sector, represents an essential effort to overcome the high costs and limited capacities of the country's road and rail transport systems and thus to boost its international competitiveness and foster sustainable economic progress (Bajas, 2010).

To a large extent, developing economies, like India, relied upon exports of agricultural and non-high value, labor intensive, manu-factured products in sustaining their economic development. The international trade of such products was heavily influenced by price and quality, of which the values and competitiveness of such products within the global market are strongly influenced by value added activities, e.g. grading, sorting, packaging, labeling, marking, refrigerating, processing, distributing, and retailing. Such require-ments, together with the development of multimodal supply chains, gradually triggered the importance of dry ports, where they often facilitated market development, seamless integration and closer collaboration between different participants of supply chains.

There were about 200 European and 370 US dry ports (or inland terminals) with varying capacities, where governments enthusiasti-cally promote their development and, in some cases, collaborated with private firms (United Nations Economic and Social Commission for Asia and the Pacific, or UNESCAP, 2006; Rahimi et al., 2008). At the same time, by 2012, about 200 dry ports had been established in India, and this number accelerated quickly, especially in view of the proposed implementation of establishing special economic zones throughout the country and the simplification of customs procedures, notably digital documentation which would enhance transparency and simplify documentation processing (Raghuram, 2005). In spite of this, further development in transport infra- and superstructures by way of capacity augmentation, mechanization and automation was imperative to realize the true potential of containerization in India which was expected to triple in between 2010 and 2020, where hinterland potential for container traffic was estimated to increase more than 70 percent.

In fact, one should not overlook the recent development of intensified competition between different ports which contributed to the blurring of boundary lines demarcating seaport's, specific hinterlands and forelands (Heaver, 2002). In other words, the locational decisions of dry ports have significant bearings on the efficiency and competitiveness of the whole supply chain, which in turn also affected the competitiveness, and even survival, of ports. However, while there has been much research investigating this topic in advanced, Western economies (for instance, Rutten, 1998; Slack, 1999; Hesse and Rodrigue, 2004; Notteboom and Rodrigue, 2005; Rahimi et al., 2008), so far, studies on developing economies remain scarce. Hence, it is very important to know more about the experiences from developing economies in order to understand the relationship between ports and dry ports.

In many Western, advanced economies, the philosophy behind the establishment of dry ports is clear. Such ports serve either as a mechanism to relieve port congestion (Slack, 1999; Roso, 2008; Roso et al., 2009) – as being part of the development process in complementing port regionalization, facilitating inland distribution and better control of the supply chains (Hesse and Rodrigue, 2004; Notteboom and Rodrigue, 2005) – or as part of the migration process of the logistics industries providing value added services (Islam et al., 2005; Notteboom and Rodrigue, 2007), from urban to more remote areas due to expensive land prices and land use restrictions (Chapman and Walker, 1991; Ducruet, 2007) including environmental and social issues. Thus, the establishment of dry ports was highly specific and predominantly demand-driven; they needed to provide high-end, value added activities. Apart from simple centers for the consolidation of cargoes, they were required to provide high-end, value-added services, while they had to be situated in intermediate locations which were close to both ports and production centers with highly efficient transport connections between different nodes.

On the other hand, dry ports or inland terminals in developing economies were usually export-oriented (rather than import-oriented). In other words, they were not just initiated by ports, but also as catalysts in the promotion of exports, regional development and inter-regional trade (UNESCAP, 2006). As mentioned earlier, dry ports often carried out functions which could also be carried out by ports, and such an 'inside-out' approach (Ng and Cetin, 2012) in the establishment of dry ports might pose some serious questions on whether they should be located inland, or surrounding the port areas. Also, under this situation, ports and dry ports might compete with each other so as to gain benefits and competitiveness from the logistical and value added activities along the supply chains. This issue will be further discussed in Chapter 8.

7.2 Dry port policies: dilemmas and the rationale for dualism

In the past two decades, more than 200 dry ports have been established at different places in India. About 40 of them were located near to the major gateway ports, with Jawaharlal Nehru Port Trust (JNPT) in Mumbai, Mundra, and Chennai being notable examples. In this regard, trucks and rail carried 58 percent and 42 percent of the port-dry port traffic, respectively (Hariharan, 2004). Until recently, most dry ports were public entities under the Indian government, notably state-owned corporations. However, the uneven distribution of dry ports within the country, with about 40 percent, 30 percent and 20 percent being located within the southern, western and northern regions respectively (UNESCAP, 2006) has led to congestion at some facilities and the breakdown of infrastructures on the one hand, alongside capacity underutilization on the other. Also, given scarce financial resources, technological and management know-how, dry ports in India were never very innovative, where long-term efficiency-enhancing investments, research and development were rarely considered; dry ports were further hindered by the government's labor protective policies. Indeed, the almost complete monopoly of state-owned corporations, notably Container Corporation of India Ltd. (CONCOR) and Central Warehousing Corporation (CWC), contributed to the above problems, especially since, as government-approved monopolies, different dry ports often provided generic solutions to non-standardized demands between different regions, raising the query on whether dry ports were really providing customer-oriented services (UNESCAP, 2005).

The major problems were typified by over regulation, poor quality service levels, inadequate infrastructure investments and underutilization. In turn, this affected the competitiveness of Indian manufactured products in the international market. According to the World Bank's Logistics Performance Index (LPI), India was ranked 39th, 47th and 46th in terms of overall performance in 2007, 2010 and 2012, respectively (World Bank, 2007, 2010a, 2012). Such relative inefficiency was largely due to the reluctance of dry port operators to offer time bound commitments to cargo owners and shipping lines, resulting in the inability of the latter to plan connections of the hinterland containers to specific ships. Unsurprisingly, these factors also led to the poor perception of dry ports (and logistics) by the public. According to anecdotal information, in India, the logistics industry, including dry ports, was perceived as 'a backward' and 'bleak' sector, thus making it difficult to attract talent, nor has it been able to impart necessary skills and vision.

To address these problems, the Indian government embarked upon a capacity enhancement program, and loosened its control on dry port operation through private participation mainly through the sale and/or leasing of facilities, joint ventures and/or BOT arrangements, with the government only maintaining regulatory functions while leaving the operational and management aspects to private operators (Haralambides and Behrens, 2000). An inter-ministerial committee for approval of applications for dry ports was established so as to facilitate single window mandatory clearances, payments, incentives, certifications and customs presence. Responding to this initiative, together with the projected container trade growth of 15 percent per annum within the next decade (Investment Commission of the Government of India, 2006), a number of dry port users, including multinational logistics service providers (e.g. Schenkers, Kuhne & Nagel) and shipping lines (e.g., APL, Maersk, etc.), joined India's dry port sector[1]. Apart from enhancing the efficiency of the supply chain, increasing foreign income (like land rents and taxes) and proving for the transfer of technology and know-how, the government anticipated that the participation of foreign firms in the operation and management of dry ports could improve the abysmal condition of India's transport infrastructure, as poor communication and transportation infrastructure could tarnish India's image for potential investors in a tangible way (Sachs et al., 2000). Here it should be noted, however, that state-owned firms, notably CONCOR, were still operating more than 60 percent of India's dry ports, resulting in a relative lack of competition within this sector.

Unsurprisingly, such a lack of competition within the sector provided little pressure to improve the rather mediocre performance of the dry ports. Nevertheless, given the close relation between transport and economic development, poor performance affected the competitiveness of Indian products overseas due to higher costs. On the one hand, the encouragement of private, especially foreign, investments was to address such an imbalance through enhancing the quality of the dry port sector, with the attempt to boost the quality of the supply chain, and thus the competitiveness of Indian manufactured products in the global market. On the other hand, foreign participation, often with superior technology, marketing strategies, management know-how and the willingness to provide time-bounded guarantees to cargo owners and shipping lines, posed significant threats to the survival of local, state-owned and operated, dry ports, especially if they had lost the advantages of the government protection umbrella. Moreover, foreign investors entering the sector usually had captive cargoes, and through the control of dry ports and inland logistics and transportation, they were perceived to generate synergetic benefits leading to significant competitive advantages. As a result, local, state-owned, dry port operators often found it difficult to survive. Moreover, given the large quantity of dry ports established around the country (and the number of employees being employed), it was politically controversial to the Indian government to allow their dry ports to be edged out by new, foreign competitors. Hence, the Indian government was trapped in a dilemma, where they would like foreign investors to assist the country's development, especially improving transport infrastructure and efficiency, but simultaneously wanted to protect local interests from being exploited by these investors who could abuse their market power (Sachs et al., 2000a). The response to such a dilemma was the so-called dualistic policies imposed by the Indian government. In this regard, three initiatives were implemented, namely, land pricing and distribution, operation, and connectivity.

7.2.1 Policies on land pricing and distribution

The major component of the capital costs involved in the construction of a dry port was the cost of land; land pricing and distribution was probably the Indian government's most important policy in dry port development. State-owned corporations, notably CONCOR, were often given preferential treatment so as to ensure that their new terminals could develop with minimal financial difficulties. According to anecdotal information, a typical dry port in India (with an annual container throughput of about 120,000 TEUs) would require about 30 acres. Land cost varied based on several factors, e.g. geographical location, availability of usable lands, economic environment, competition with other potential land users, proximity to market place or gateway ports, and close affinity to road/rail networks. Taking the above factors into consideration, the land price varied between 50,000 to 150,000 US dollars per acre, and thus a dry port with an annual throughput of 120,000 TEUs would typically need to spend 1.5–4.5 million US dollars just to acquire the required land for building the terminals and installing the infra- and superstructures. Moreover, being the largest landowner within India, the government had a major presence in its land's sale and lease. While charging market prices to foreign investors, except in areas around JNPT, lands had been leased out to state-owned firms, especially CONCOR, for very long periods (usually 99 years) at rates significantly under market values. In fact, the government itself was a price setter and implemented dual pricing on land sales or leases, despite the fact that such actions directly violated its own Competition Act (2002), Article 3, No. 1 (a), stating that no persons/associations should undertake actions which could determine prices (Government of India, 2003). According to the anecdotal information, in some cases, the financial commitments on land could even be waived completely, as long as government officials were compromising alternative arrangements to assure that their economic interests and policy objectives could be maintained. Such substantial differences in the costs of purchasing land placed the foreign investors in a seriously disadvantaged position against state-owned corporations. In this way, the Indian government was actually subsidizing state-owned corporations by substantially reducing their initial capital costs when setting up dry ports, but *vice versa* for foreign operators.

Apart from this financial factor, acquiring land for the construction of dry ports, especially green-field projects, was subject to governmental permission in changing the land use purpose (as in almost all cases, the land was initially used for agricultural purposes), and the land distribution policy was clearly offering preferential treatment to state-owned corporations which often enjoyed the privilege of obtaining the required land against other private operators. For example, in Mumbai, while most of the dry ports surrounding JNPT were privately owned and operated, until very recently, the Indian government had only granted land to the dry port operated by CONCOR in the construction of railhead, leaving all others to transport their containers by trucks. Given that the typical cargo transportation rate between dry and gateway ports cost about 0.15 and 0.25 US dollars per km by trains and trucks respectively (Ng and Gujar, 2009), the government had clearly imposed preferential policies so as to trigger the attractiveness of state-owned corporations against its competitors. The acquirement of land was further complicated by the non-availability of proper land records, thus often leading to the wastage of time and litigation. To resolve such problems, the Indian government, through local bodies like the City Industrial Development Corporation (CIDCO) or other similar bodies, acquired land from users and developed it after which it was sold or leased to the interested parties. However this process was often riddled with corruption. According to anecdotal information, it was common practice that the promised road, electricity, telecom infrastructures were never delivered or delayed due to the fact that government officials, especially local ones, were often waiting for 'credits' before taking the initiative in acquiring the required land.

7.2.2   Policies on operation

Apart from land policy, the Indian government carried out policies allowing dry ports operated by state-owned corporations to grow, mature and compete. State-owned dry ports were allowed to suffer financial losses for the initial two years. This largely benefited such state-owned corporations as it implied that a guarantee was offered by the government to absorb any losses which were incurred during this period. In fact, even if the objective of balancing the books failed after that, the operators could cut their losses through selling off their equity holdings, either partially or fully. Also, apart from such breathing space, in some cases, the Indian government would also take up the responsibility of partially covering revenue costs which were mainly comprised of the transport costs between dry ports and ports. Such assistance often enabled state-owned dry ports to use discounts, preferential and predatory pricing to attract users. According to anecdotal information, users usually enjoyed fuel costs offered at half price when they called state-owned dry ports. Of course, this significantly enhanced their competitiveness, especially considering the substantial increase in oil prices in the past decade. Perhaps equally important was that, such a subsidizing policy implied that rail could be employed even if the threshold was not fulfilled. In India, it was generally understood that a minimum demand of 90 TEUs per day were required for freight rail to become economically feasible (Ng and Gujar, 2009) (Illustration 7.1). In this case, even CONCOR officials admitted that, except during the congested seasons, the real prices were often flexible and significantly lower than what had been stated publicly, especially towards their major users, where bulk discounts, extended credit periods and storage offered at subsidized rates were common, even despite the enactment of the Competition Act (2002). In certain cases where the cargoes involved were time sensitive or prone to pilferage, e.g. perishable products, garments and accessories, household products, leather products, and pharmaceuticals, the operators often demanded a premium above the published tariff simply whilst handling such cargoes. In fact, it was also not uncommon for the operators to rent out, partially or entirely, the storage space to a single user for certain time periods against payment in advance, which relieved substantial cash flow burdens.

Illustration 7.1   A minimum demand threshold is required for freight rail to become economically feasible in India

Source: Authors, taken in Delhi, India (2009).

7.2.3   Policies on connectivity

Dualistic policies also affected the connectivity between dry ports and gateway ports. In this regard, the influence on rail operation served as the most important example reflecting such a policy. Before 2006, domestic container transport was fully reserved to CONCOR, but the government decided to invite private participation into the container rail sector (World Bank, 2002), for two major reasons. First, the rapid growth rate in container trade implied an additional container transportation of almost one million TEUs annually, and hence it was perceived that CONCOR alone could not handle this increase. Also, the government wanted to ease the congestion on road transport (nearly 60 percent) by transferring some of the burden to rail. The result was the granting of licenses to 14 foreign private firms to operate container trains between dry ports and the gateway ports through concession agreements envisaging two categories, namely: (i) pan-India basis, costing one million US dollars (chosen by ten operators); and (ii) the more expensive (but more lucrative) route-specific basis, costing 2.5 million US dollars (chosen by four operators). Through the same agreement, private operators would deploy their own containers, wagons and handling equipment, building their own terminals, and marketing for customers. Under such arrangements, the participation of the private sector was expected not only to attract traffic from road, but also enhanced rail's capacity without posing extra financial burdens to the Indian government.

Nevertheless, after acquiring expensive licenses, foreign operators often found it difficult to sustain their operations as these expenses adversely affected their initial pricing structure, not to mention the substantial capital costs of land purchase for terminal construction (as discussed earlier). To make things worse, they also grudged the absence of service level guarantees in the concession agreements which inhibited their abilities to attract cargo from road. Private operators were further disadvantaged because they often lacked their own terminals and experienced a shortage of container wagons[2]; this meant that they often needed to pay CONCOR to use their terminals and container wagons, as a result the private operators found that their supply of services was (nearly) always prone to the government's manipulation. Transport costs were highly subsidized by the government, usually in terms of fuel subsidies, as well as rail and road haulage between state-owned dry ports and the gateway seaports and production plants, as exemplified by the fact that the construction of CONCOR's railheads were heavily funded by public money. Under this situation, the private operators soon found that the carrots which initially attracted them, i.e. the projected substantial container trade growth, had evaporated quickly.

7.3   Impacts on the spatial characteristics of dry ports: three case studies

Understanding the dualistic approach of the Indian government policies, three major industrial clusters, located in different parts of India, were studied by the authors. The locations of the three studied cases can be found in Figure 7.1. Note that all of the studied dry ports were closely linked to their respective ports which facilitated the export of agricultural and manufactured products.

Figure 7.1   The locations of the three studied cases in India
Source: Authors.

The three studied cases were investigated by applying the grid technique. Indeed, a foremost concern of spatial analysis was the 'friction of distance,' i.e. impediments to movement, occurring due to spatial separation, which often involved economic and/or financial costs. The technique was a heuristic approach in determining the optimal location of fixed facilities (in this case, dry ports) based on the least-cost center for moving in- or outbound cargoes within the geographical grid concerned. It assumed that the originating sources and outbound destinations for in- and outbound cargoes respectively were fixed, and that the (dry port) operators had concrete ideas on the approximate volumes of cargoes that they were likely to handle. Also, it integrated both spatial and non-spatial data in solving problems in transport engineering, with the shortest path analysis being a precursor to this technique. In other words, the simulated optimal solution would be the place with minimum transport costs. A detailed explanation on the grid technique, including the mathematical formulations, can be found in the appendix at the end of this chapter. During the analysis, several assumptions were made. First, there would be no significant variations between different dry ports in terms of efficiency. Second, the unit transport cost had a linear relation with distance. Third, unacceptable or inaccessible routes did not exist. Fourth, only local cargoes (within 100 km from the production bases) were considered. Fifth, not calling a dry port was not an option. As mentioned earlier, dry ports were more than just load and distribution centers which could also serve additional necessary functions in facilitating the shipment process along the supply chain. Moreover, given that Indian shippers largely consisted of medium- and small-sized firms, it was practically impossible for most of them to get around dry ports and ship their cargoes to the gateway ports directly. Sixth, the analysis was based on existing transport infrastructure and facilities. Seventh, only one dry port would be called each time. Eighth, the simulation was based on a single-facility location model. Finally, it was assumed that rail, instead of trucks, would be used, as long as the route concerned could fulfill the following criteria: (i) the annual cargo size along this route reached a minimum threshold of 32,400 TEUs (90 TEUs per day; 360 days per year); and (ii) this route was actually supported by rail to ports. The data necessary for model simulation was collected from various industrial sources, including data available on the official website of different dry ports (Table 7.1), as well as the field trip to India undertaken by the authors as mentioned before.

To offer a clearer picture on the choice of dry ports by shippers, a number of existing dry ports were also included in Table 7.1, namely Kudalnagar, Bhusawal, Daulatabad, Ankleshwar and Gandhidham Inland Container Depots (ICDs)/Container Freight Stations (CFS). All of them shared common characteristics, i.e. they were closely located ($\leq$ 20 km) from the simulated optimal dry port locations of studied cases. With such understanding, it meant that under the current situation, most of the cargoes generated from the production bases

Table 7.1 The paved areas, capacities and throughputs of selected dry ports in India

Dry port	Paved area in 2012 (m²)	Capacity in 2012 (TEUs)	Throughputs (TEUs)	
			2006–2007	2010–2011
Southern India – Tirupur				
Tirupur ICD (TICD)	1,553	14,335	3,795	2,034
Kudalnagar ICD, Madurai (KICD)	8,580	79,200	438	8
Central India – Nagpur				
Nagpur ICD (NICD)	67,750	625,390	75,452	72,814
Bhusawal ICD (BICD)	20,230	186,700	2,534	13,802
Daulatabad ICD (DICD)	12,576	116,100	5,774	16,415
Northern India – Ahmadabad				
Sabarmati ICD (SICD)	128,428	1,185,500	112,616	147,637
Ankleshwar ICD (AICD)	86,978	802,881	1,568	16,993
Gandhidham CFS (GCFS)	121,406	1,120,700	4,032	15,813

Notes: Remarks: Capacity is calculated using the following formula: (paved area)*(3 layers stacked containers)*(50 weeks of operations)÷(the area occupied by 1 TEU). Source for paved areas and throughputs: CONCOR website.

were exported via their respective local dry ports, i.e. TICD, NICD and SICD.

7.3.1 Case one: Southern India

With a population of 400,000 spreading over 30 km², Tirupur was located in the state of Tamilnadu. Known for its textile industries, the city generated apparel exports worth 1.5 billion US dollars annually, equivalent to nearly 40 percent of India's total garment export value. There were about 3,000 knitting, stitching, dyeing, bleaching and printing units in the region manufacturing garments which exported mainly to Europe and the US. Almost all the cargoes were exported by sea, mainly through the ports of Tuticorin and Cochin. The city's local dry port, TICD, commissioned in January

2005 and operated by CONCOR, spreading over 0.7 hectares, was located 7 km away from the core production bases. TICD had a covered warehouse measuring 300 m² with a custom clearance facility. When this study took place, however, TICD was not connected by rail tracks to any of the ports and all cargoes had to be carried by trucks, and neither the national nor the Tamilnadu state government had any concrete plans to construct railway lines connecting TICD and the ports. Apart from TICD, a small amount of cargoes would have to be cleared at KICD located at Madurai. The current and simulated solutions of Southern India (Tirupur) can be graphically represented in Figure 7.2.

Current Solution: [Tirupur] → [TICD/KICD] →[Cochin/Tuticorin]
Simulated Solution: [Tirupur] → [Optimal Dry Port] →[Cochin/ Tuticorin]

By applying the grid technique, the optimal location for Southern India (Tirupur) could be visualized in Figure 7.3.[3] As per simulated results, the optimal location of the dry port in serving Southern India (Tirupur) should be near Madurai which was approximately midway between the production bases and the ports. This location was 105 km away south from TICD (located only 20 km from Tirupur's major production base).

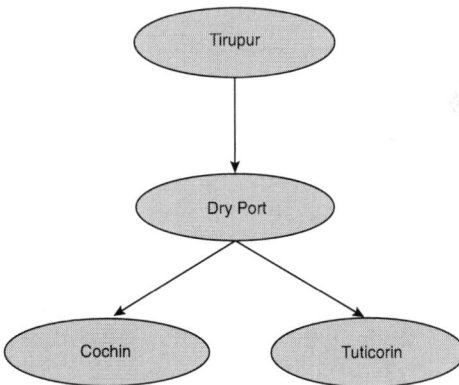

Figure 7.2 A graphical representation of the situation of the case study in Southern India

Remark: arcs were not constructed in accordance to actual distances.
Source: Authors.

Figure 7.3   The optimal dry port's location for Southern India
Source: Authors.

7.3.2   Case two: Central India

Nagpur was an old city located in the state of Maharashtra, with a population of three million spreading over 40 km². It was located in a region rich in forestry and mineral resources. Hence, the major industries were mainly agricultural and mineral (or directly related) products, e.g. cotton, soya, rayon, paper, iron/steel, aluminum, and the like. Its local dry port, NICD, was located 12 km away from the major production bases. Despite the fact that the port of Vishakhapatnam was equidistant from Nagpur (and also connected by railroads), nearly all cargoes from Nagpur (according to industrial information, about 98 percent) were shipped out through JNPT, which was also connected with NICD by road and railroads. The location of NICD can be found in Figure 7.5. Apart from NICD, a small amount of cargoes was cleared at BICD and DICD, both located within Maharashtra, approximately midway between Nagpur and JNPT. The current and simulated solutions of Central India (Nagpur) can be graphically represented in Figure 7.4.

Current solution: [Nagpur] → [NICD/BICD/DICD] → [JNPT]
Simulated solution: [Nagpur] → [Optimal Dry Port] →[JNPT]

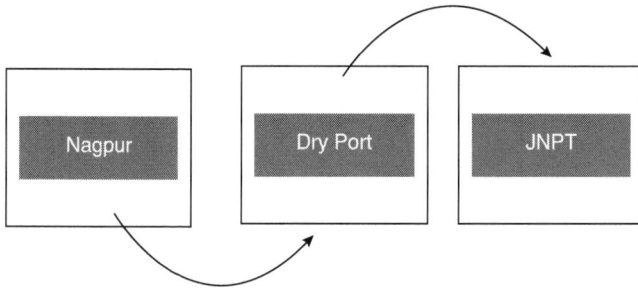

Figure 7.4 A graphical representation of the situation of the case study in Central India

Remark: arcs are not constructed in accordance to actual distances.
Source: Authors.

Figure 7.5 The optimal dry port's location for Central India
Source: Authors.

By applying the grid technique, the optimal location for Central India (Nagpur) can be visualized in Figure 7.5. As per the simulated results, the optimal location for a dry port in serving Central India (Nagpur) should be 150 km to the southwest of Nagpur's production base, towards the direction of JNPT.

7.3.3   Case three: Northern India

With a population of five million spreading over 50 km^2, Ahmadabad was located in Northern India and was the capital of the state of

Gujarat. It was famous for its textile mills dating back to the last century. There were also other industries, notably pharmaceuticals, paper, sheet glass, chemicals, and agricultural products, like oil cake and edible oil. Its local dry port, SICD, was located about four km from its major production bases spreading over ten hectares and was connected by road and rail to the ports of JNPT, Mundra and Pipavav. According to anecdotal information, 67 percent, 20 percent and 13 percent of the cargoes were shipped out through the ports of JNPT, Mundra and Pipavav respectively. Apart from SICD, a small amount of cargoes were also cleared at AICD and GCFS, both located within Gujarat. The current and simulated solutions of Northern India (Ahmadabad) could be graphically represented in Figure 7.6.

Current solution: [Ahmadabad] → [SICD/AICD/GCFS] → [JNPT/ Mundra/Pipavav]

Simulated solution: [Ahmadabad] → [Optimal Dry Port] →[JNPT/ Mundra/Pipavav]

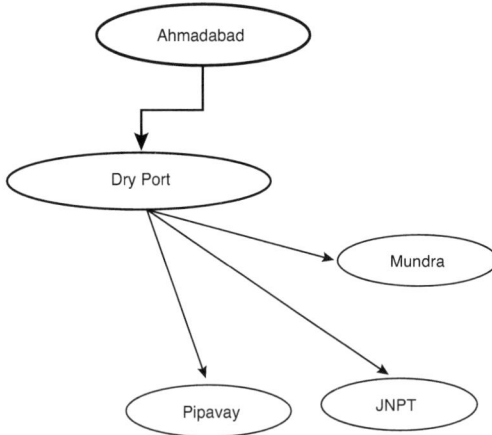

Figure 7.6 A graphical representation of the situation of the case study in Northern India

Remark: arcs are not constructed in accordance to actual distances.
Source: Authors.

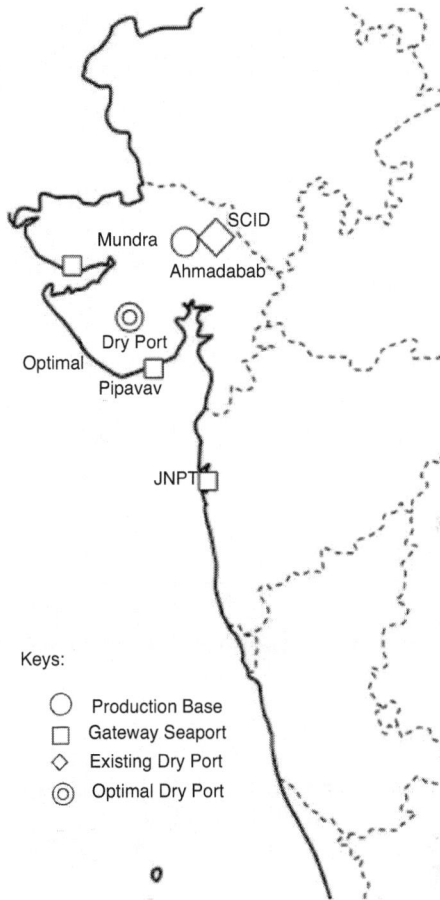

Figure 7.7 The optimal dry port's location for Northern India
Source: Authors.

By applying the grid technique, the optimal location for Northern India (Ahmadabad) is visualized in Figure 7.7. As per the simulated results, the optimal location of a dry port in serving Northern India (Ahmadabad) should be approximately 170 km to the southwest of Ahmadabad's production base, which is significantly more proximate to the major gateway seaports of JNPT, Mundra and Pipavav.

7.4   Discussion and conclusion

All the simulated optimal dry ports shared common locational characteristics, of which they were located in sites with significant distances away from both the production bases and gateway ports. As per the simulated results, existing dry ports which were located proximate to the optimal dry ports, i.e. Kudalnagar, Bhusawal, Daulatabad, Ankleshwar and Gandhidham ICD/CFS for Tirupur, Nagpur and Ahmadabad respectively, should possess the best potential in attracting the most cargoes from the production bases. However, as indicated by the distribution of cargoes between different dry ports, they were significantly different from the real situation where local dry ports, i.e. TICD, SICD and NICD, respectively, had significantly higher throughputs than their counterparts. It was clear that all the simulated optimal locations failed to reflect the reality. In other words, it meant that only dry ports withlocations proximate to the production bases could attract cargoes of any significance (TICD, SICD and NICD were only located 7, 12 and 4 km away from their respective production bases).

In this regard, an anomaly existed where the pulling force of centrality was more influential than intermediacy (cf. Fleming and Hayuth, 1994; Behrans et al., 2007), supporting the notion that the impact of different types of costs on the locational choice of facility users could be diversified in the decision of shippers in using dry ports. In India, shippers clearly sacrificed transport cost-savings in return for alternative benefits, say, convenience, relation, better control, to name but a few options. Indeed, the three studied cases illustrated that convenience, local relationship and better local control often served as equally, if not more, important considerations on deciding whether dry ports, and their facilities, would become successful. It also indicated that when the spatial pattern of economic masses was uneven, dry ports tended to locate very close to the dominant economic mass (or a group of economic masses) that were geographically proximate. This led to a suspicion that the normative, least-cost approach was an insufficient specification, which is in stark contrast to many experiences from Western, advanced economies where dry ports were often located at intermediate sites along the intersections of transport corridors (Rahimi et al., 2008).

Hence, the analysis in this chapter suggests that the competitive structure of the Indian dry port industry did not only have economic, but also the influence of *environment heterogeneity* (see Chapter 5) in affecting the final settings. This was in line with de Langen and Pallis (2007) where partial lowering of entry barriers had often been the rule of the game in organizational reforms where, similar to port reform and governance, local and regional interests and political culture were equally, if not more, important (Kingdon, 1995; Ng and Pallis, 2010) in deciding the rules of the game, especially when reservations to changes and foreign participation were still prevalent. In this regard, one should not overlook the fact that the Indian government undertook policies to assist foreign firms in establishing themselves within the sector[4], nor that local dry port operators made no attempts to improve themselves, as witnessed by CONCOR's attempts in providing value added services, e.g. packaging, container repairs, and storage (Sahay and Mohan, 2003).[5] However, it was the government which tolerated (and encouraged) foreign investments in dry ports so as to exert pressure to improve the sector's factor conditions but, simultaneously, it was also the same government which intervened so as to slow down the need for state-owned corporations to improve themselves while also sustaining their dominating position, so that foreign investors would not be given any opportunities to create (and substantially benefit from) any market concentration of power. As Ng and Tongzon (2010) pointed out, largely due to the uncertain environment, dry port operators were discouraged from long term planning and investments so as to expand service areas, and this impeded the logistical roles of dry ports in terms of promoting trade and interactions among different regions, and prevented them from fully integrating into the global supply chains. In fact, a similar problem exists among Brazilian dry ports, which will be further discussed in Chapter 8.

According to anecdotal information, an important reason for foreign firms to continue their dry port operations, despite often operating at a loss due to dualism[6] was to minimize the risks due to mediocre performance along the supply chains. Also, such institutional barriers slowed down the development, and integration, of ports and dry ports into supply chains, thus contributing to rather poor logistical performance as mentioned earlier. According to various interviewees, the inability of integrating gateway ports, dry ports and other inland logistical components boosted the need for interpersonal flexibility and better control. In India, although the number of shippers was fairly large, the average quantity of cargoes exported by each was actually quite small, thus consolidation was always required so as to fill up the containers. Hence, as indicated in the studied cases, with the relative lack of efficient logistics, shippers often preferred local dry ports so as to obtain better information and found it easier to tackle upcoming challenges which required quick or immediate responses. In this regard, Schoenberger (1997) further highlighted the importance of interpersonal flexibility, especially with increased uncertainties arising from technological changes and global competition. With substantial uncertainties generated by rather inefficient logistical performance, shippers felt more secure when they retained control over their consignment until the last moment, which simultaneously enabled them to negotiate better freight rates. Unsurprisingly, they established much closer relations with personnel within local dry ports, which often saved substantial resources and allowed them to gain important knowledge through the reduction of uncertainties due to shared culture, norms, attitudes, values and expectations (Gertler, 2001; Vicente and Suire, 2007). The consequence was the generation of trust in documentation clearance, custom inspections and preferential treatment (Illustration 7.2). This strongly suggested that local dry ports, i.e. TICD, SICD and NICD, were auxiliary sectors forming parts of the agglomeration process of respective production bases, where the latter underwent spatial clustering and gradually evolved into central places, especially given the relative small size of shippers, that were unable to internalize services necessary for exports (Webber, 1972). Non-local dry ports, on the contrary, could not benefit from this agglomeration process due to their relative lack of connections with major production bases.

The specific demands for dry ports in the three studied cases were clearly lacking, where dry ports were simply regarded as load centers rather than integrated components of supply chains. As mentioned, the Indian government also expected dry ports to play catalytic roles in stimulating economic development of the surrounding areas (UNESCAP, 2006), and hence their establishment was based on the perception that dry ports could be self-sustaining and brought in

Illustration 7.2 The importance of developing trust and relationship between shippers and custom officers is explicit in India

Source: Authors, taken in Delhi, India (2009).

perpetuity effects (and stimulated demands) to surrounding regions. This philosophy largely explained the major motivation behind the large number of dry ports being established throughout the country in the past decade without the guarantee of genuine demands, nor clear roles and functions, thus contributing to their high degrees of similarities and immobility, not to mention their inabilities in providing value-added services to shippers. Such a lack of demand, the high degree of similarities and the immobility of dry ports paralyzed the competitive strength of those which were not proximate to the production bases, as they could not improve their attractiveness through spatial re-arrangement or product differentials (Hotelling, 1929; Hoover and Giarratani, 1985) in face of the non-financial benefits that shippers could enjoy by using local dry ports.

The inability of integrating dry ports into supply chains was made more difficult by the aforementioned existence of abundant, but mainly small- and medium-sized, shippers within India (Illustration 7.3). Given their size and the nature of their manufactured merchandizes, the prerequisites in stimulating the growth of value added activities, or integrating the key business processes from end users through to original suppliers, as required by supply chains nowadays, hardly existed. Unsurprisingly, shippers within the studied cases found it very difficult (if they ever tried in the first place) to play more innovative roles to build up the centrality of dry ports, thus further raising the difficulties in overturning the status quo. Hence, the view that installing dry ports at carefully chosen locations could automatically enable them to evolve from simple warehousing functions to logistical centers seemed to be an inaccurate assessment that exhibited a lack of understanding of the real situation. This further discouraged any logistical and value added activities to move further away from either the port areas or the major production bases. Simply speaking, while Indian dry ports were able to carry out certain bureaucratic functions (only when located near the port areas or major production bases), they failed to carry out the equally important logistical functions which contributed to the integration, and thus efficiency and effectiveness, of supply chains. Nevertheless, the three studied cases also illustrated

Illustration 7.3 In India, dry ports played very important roles in consolidating many small quantity of cargoes into containers

Source: Authors, taken in Delhi, India (2009).

that dry ports, and their facilities, either located around the gateway ports or inland production bases, could both become successful. This provided further evidence explaining the inadequacy of port-centric logistics and the port's strategy of inland penetration in developing economies, as well as their complementary nature.

Also, the inability of non-local dry ports to attract business confirmed the proposition that the existence of infrastructure and other facilities by themselves were unlikely to stimulate regional development. Hence, rather than being 'supply chain-oriented,' as exemplified by the studied cases, dry ports in developing economies like India seemed to be more 'cluster-oriented,' reflecting their dynamic interrelation with the industrial clusters surrounding them. These results also supported Hall's arguments, where diversifications existed through a process of regional institutional transformation (Hall, 2003). In terms of policy implications, the studied cases supported the proposition that the Indian government's strategy of stimulating regional development by establishing simple, generic dry ports around the country largely failed to live up expectations. To enable them to become the catalysts for regional economic development, governments should undertake initiatives to enhance their attractiveness by stimulating demands for such facilities and defining the objectives and roles for each dry port, rather than just the physical installation of similar facilities. To minimize the wastage of resources, when deciding whether and where dry ports should be established, decision-makers should first address an important question – whether their country possessed the necessary conditions in establishing supply chain- or cluster-oriented dry ports. Moreover, the dualistic policies of the Indian government on the dry port sector also failed to stimulate any genuine demands for the use of dry ports. Simultaneously, they discouraged foreign investors, which usually possessed captive cargoes as mentioned earlier, to help in stimulating demands and to further develop integrated supply chains after the installation of infrastructures and facilities.

Last but not least, the analysis in this chapter also highlighted the importance of institutions and policies in promoting (or hindering) the successful development of port-included integrated transportation and logistical systems. As noted by Gertler (2001) and Ng and Pallis (2010), specific backgrounds, culture and local interests could lead to different interpretations of even the same concept, and thus warn against the attempt of imposing generic solutions to different regions even when addressing a similar problem. This also confirmed the proposition that the spatial patterns between different types of economies could be diversified due to fundamental geographical differences and the pace of regional development. It seemed sensible to argue that the agglomeration of logistical and value-added activities around port areas and major production bases should be an appropriate way forward for developing economies to strengthen their supply chains, and gateway ports and dry ports should be assisted by their respective governments so as to allow them to develop into hubs which could provide logistical and value added services effectively. At the same time, it further supported the port-focal logistics concept as proposed by the authors earlier in this book (see Chapter 1). This might help them to achieve the threshold and centrality (Fleming and Hayuth, 1994) to stimulate demands for further enhancement in supply chain quality and regional development.

In the next chapter, the authors will bring readers to another developing economy, namely Brazil. Although located on the other side of the world, it was surprising to find that the development of Brazilian ports and dry ports, including their interrelationship and the degree of integration into supply chains, shared a lot of similarities with the Indian situation that we just reviewed and analyzed.

Appendix: The Grid Technique

The grid technique superimposes a grid upon a geographical area containing particular cargoes from originating sources to the final destinations. The grid's zero point corresponds to an exact geographic location, as do the other points. Every source and destination can then be determined by the grid's coordinate. It defines each source and destination location in terms of its horizontal and vertical grid coordinates. In this regard, it is possible to visualize the technique's underlying concept as a series of strings, i.e. attached to weights corresponding to the weights of in-and outbound cargoes which, in this case, dry port operator handle.

Here it is important to note that the application of the grid technique is based on the normative view of location where: (i) land is isotropic and uniform in resource ability without any significant barriers to movements; (ii) population is uniform in all respects; and (iii) perfectly competitive markets exist, in which producers and consumers possess perfect market knowledge. The grid technique can be expressed as:

$$C_{(x,y)} = \frac{\sum(r*d*S) + \sum(R*D*M)}{\sum(r*S) + \sum(R*M)} \qquad (1)$$

s.t.

$C,M,S,r,d \geq 0$

where C is the 'center of mass,' i.e. the optimal location, D is the distance from 0 point on the grid to the grid location of outbound cargoes, d is the distance from 0 point on the grid to the grid location of inbound cargoes, M is the weight (volume) of outbound cargoes, S is the volume of inbound cargoes, R is the outbound cargo transportation rate/distance unit for the cargo and r is the inbound cargo transportation rate/distance unit for the cargoes. R and r are the transport rates per distance unit. To determine the least cost center on the grid, it is necessary to compute two grid coordinates, one for moving the commodities along the horizontal axis and one for moving them along the vertical axis. Both coordinates are computed by using the grid technique formula for each direction.

8
Case Study – Brazil

This chapter discusses how institutional factors affected the bureaucratic and logistic roles of dry ports in Brazil focusing on its three states, namely Sao Paulo, Minas Gerais and Goias. They provided sufficient context for this purpose, with Sao Paulo being Brazil's economic powerhouse where 25 dry ports were located, Minas Gerais had a strong presence in the agricultural and mining-metallurgical sectors, where five dry ports were located, and Goias was an agricultural state located deep inland where two dry ports were located (Figure 8.1 and Table 8.1). In this chapter, the authors investigated the evolution of the regulatory process, the roles of institutions and how changes affected the behavior of key stakeholders and the development of dry ports. By linking institutional arrangements and the roles of dry ports, it was argued that the institutional system in place – either with reference to the entire nation and polity, or to particular related arrangements regarding the setting of dry ports – acted as the *environment heterogeneity* in boosting the enhanced bureaucratic functions of dry ports, while paralyzing their logistical functions.

The study was undertaken through the assessment of major legal documents and governmental reports. Also, the authors conducted 30 semi-structured, in-depth interviews with stakeholders (hereinafter called 'interviewees') from the three studied states. All the interviewees possessed extensive professional experience within the industries. Also, when the interviews took place, they were authorized to make key decisions for their respective organizations. Interviewees

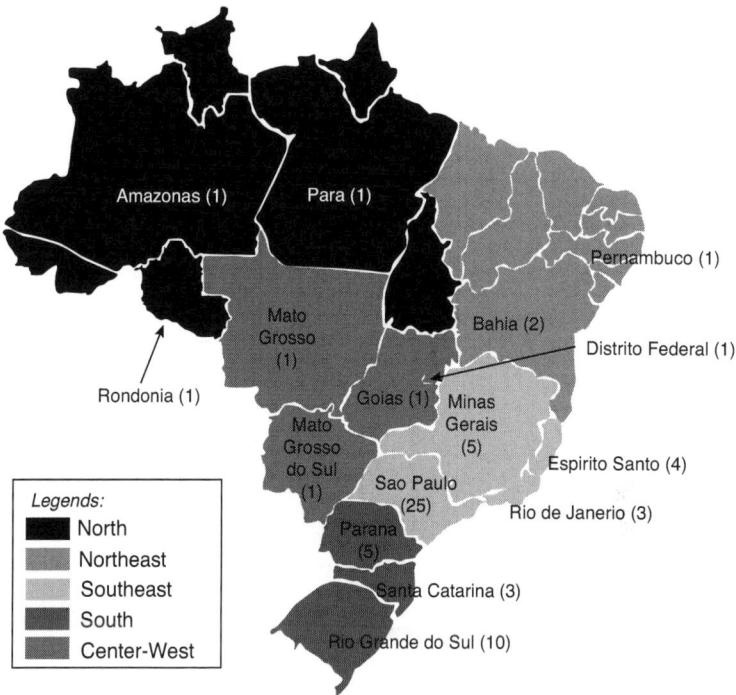

Figure 8.1   The geographical distribution of dry ports in Brazil, 2012
Source: Authors.

included specialists from the Special Port Secretariat, senior representatives from *Aduaneiras, Fundação Dom Cabral*, the Brazilian Institute of Logistics (IBRALOG), the Syndicate of Customs Brokers, the Secretary of Transport of Sao Paulo, senior representatives from the dry ports in Campinas, Guarulhos, Bauru, Ribeirão Preto, Santos and in the city of Sao Paulo. The rest of the chapter is structured as follows. It will start with a discussion on how institutional limbo in the past decades has affected the relationship between ports and dry ports in Brazil, as well as their bureaucratic and logistical roles. Towards the end, a detailed analysis will be undertaken on how institutional factors have affected the spatial evolution of the port of Santos and the dry ports in the state of Sao Paulo.

Table 8.1 The list of Brazilian dry ports included in this study

State	City	Jurisdiction	Operator
Goias	Anapolis	DRF/Anápolis	Porto Seco Centro-Oeste S.A.
Minas Gerais	Betim	IRF/Belo Horizonte	EADI – Usifast Log. Industrial S/A
	Juiz de Fora	DRF/Juiz de Fora	Multiterminais Alfandegados do Brasil Ltda.
	Uberaba	DRF/Uberaba	Empresa de Transportes Líder Ltda.
	Uberlandia	DRF/Uberlândia	Mineração Andirá Ltda.
	Varginha	DRF/Varginha	Armazéns Gerais Agrícola Lta.
Rio Grande do Sul	Uruguaiana	DRF/Uruguaiana	America Latina Logistica do Brasil S.A
Sao Paulo	Barueri	ALF/São Paulo	Armazéns Gerais Columbia S/A
	Bauru	DRF/Bauru	Cia Paulista de Armaz. Gerais Ad. Exp. e Imp. S/A
	Campinas	ALF/A. I. de Viracopos	Armazéns Gerais Columbia S/A
	Campinas	ALF/A. I. de Viracopos	Libra Port Campinas S/A
	Franca	DRF/Franca	Emp. Bras. de Armaz.Gerais, Term. e Entrep.Ltda
	Guaruja	ALF/Porto de Santos	Mesquita S/A – Transporte e Serviços
	Guarulhos	ALF/São Paulo	Plan Service Despachos Aduaneiros e Trans Ltda. (Dry Port)
	Guarulhos	ALF/São Paulo	Transquadros Armazéns Alfandegados S/A
	Jacarei	DRF/São José dos Campos	Universal Armazéns Gerais e Alfandegados Ltda
	Ribeirao Preto	DRF/Ribeirão Preto	Rodrimar S/A Transportes Eq. Ind. e Arm.Gerais
	Santo André	ALF/São Paulo	EADI – Santo André Terminal de Cargas Ltda.
	Santos	ALF/Porto de Santos	Armazéns Gerais Columbia S/A
	Santos	ALF/Porto de Santos	Mesquita S/A – Transporte e Serviços
	Santos	ALF/Porto de Santos	Integral Transporte e Agenciamento Marítimo Ltda
	Santos	ALF/Porto de Santos	Deicmar S/A – Desp. Aduaneiros Assessoria Transporte

Continued

Sao Bernardo do Campo	ALF/São Paulo	Armaz. Gerais e Entrep. S. Bernardo do Campo S.A
São Bernardo do Campo	ALF/São Paulo	SBC – Integral Transporte Marítimo e Agenciamento Ltda
Sao Jose do Rio Preto	DRF/São José do Rio Preto	Automotive Distribuição e Logística Ltda.
Sao Paulo	ALF/São Paulo	Cia Nacional de Armazéns Alfandegados
Sao Paulo	ALF/São Paulo	Armazéns Gerais Columbia S/A
Sao Paulo	ALF/São Paulo	ENBRAGEN -Emp.Bras. de Arm.Gerais, Term. e Entrep.Ltda
Sao Sebastião	DRF/São Sebastião	Cia Nacional de Armazéns Alfandegados
Sorocaba	DRF/Sorocaba	Aurora Terminais e Serviços Ltda
Suzano	ALF/São Paulo	Cia Regional de Armaz Gerais e Entrep. Aduaneiros
Taubate	DRF/Taubaté	Estação Aduaneira Interior/Taubaté Ltda

Source: Authors.

8.1   Institutional reforms and the relation between ports and dry ports

Even though dry ports existed in Brazil since the 1970s, they increased in quantity and significance only after the 1990s when the country undertook significant political and economic transformation. Until the late 1980s, Brazil consisted of a highly protected economy with a very strong state apparatus dominated by political elites. With the past orthodoxies of central planning and protectionism proving to be ineffective, there was a need for fundamental economic and political reforms.

The Constitution of 1988 marked the transition of the country into democracy, with laws that would advance individual freedoms and private enterprise. This paved the way for the decentralization of power from the federal (national) government to state and municipal levels, and opened strategic sectors of the economy, such as banking and infrastructure, to foreign direct investments (FDIs). The government led by Fernando Collor de Mello (1990–1992) initiated trade liberalization with significant cuts in tariff and non-tariff barriers. Law 8.031/90 created the National Program of Divestiture (*Programa Nacional de Desestatização*) and allowed the privatization of public firms within the steel, petrochemical and fertilizer industries. Later, under Itamar Franco (1992–1994), a new currency (*real*) effectively suppressed hyperinflation and established the basis for future macroeconomic stability backed by fiscal austerity, inflation targeting policies and high interest rates. The subsequent government led by Fernando Henrique Cardoso (1995–2003) further opened several sectors, like telecommunications, coastal and domestic navigation, and the oil and gas industries to private participation.

It was the combination of increasing demands for port services (due to trade liberalization) and the chaotic situation at port terminals during this period which led to the establishment of numerous dry ports.[1] According to the Federal Revenue, by the mid-2000s, more than 60 dry ports in Brazil were established, operated by 46 private operators. Six had gone through the public tender, with another nine waiting to do so. However, while the strategic and economic importance of dry ports increased, they faced two major challenges, namely competition from ports (following the privatization of port terminals) and the lack of a clear institutional system to regulate the sector. In this regard, competition could be linked to port reform legislation which was a decisive institutional shift brought about by the introduction of new rules to govern Brazilian ports. The role of dry ports was significantly affected when the Port Modernization Act (8630) was introduced in 1993. This institutional change brought significant improvements to Brazil's port system. It decentralized the administrative structure of ports, shifting power from the federal to state and municipal governments, and introduced measures to improve governance, notably the creation of the Port Authority Council. Previously a source of inflexibility at ports, the monopoly held by the syndicate of dockworkers had been overcome, making it mandatory that port operators employed their own staff. The old syndicate was replaced by another entity, namely the Labor Management Entities (*Órgãos Gestores de Mão-de-Obra*, or OGMO), established by the port operators for human resource management. Meanwhile, various aspects of port operations, like handling, movement, inspection, repairs, security and maintenance of vessels, were taken over by private firms through lease contracts (Rodrigues, 2009).

As mentioned earlier, the dry ports were created to reduce congestion at ports at a time when state-controlled port operations were highly inefficient. With the introduction of the Port Modernization Act, institutional reforms and new investments on port capacities led to significant efficiency gains. In turn, this affected the role of dry ports as port terminal operators started to compete for business with them. To stay competitive, they needed to be more efficient and to provide more sophisticated services. The majority of dry ports faced difficulties and closure, for instance, Piracicaba the dry port in Sao Paulo, and it was clear that they needed to diversify their activities and provide more value-added services. The reformed framework initiated a change where privatization caused a competitive relationship between ports and dry ports, with ports playing a dominating role. In this highly competitive environment, unless ports were facing significant congestion, they had few incentives to facilitate cargoes making smooth passage to dry ports (where efficient supply chains were supposed to function). All the dry port interviewees that responded to the authors referred to the difficult relationship with ports, which often created barriers, or anti-competitive measures, to complicate the transfer of cargoes to/

from dry ports.[2] An illustrative example was the incidence of an additional terminal handling charge called THC2 – demanded by the port terminal operators for the release of cargoes from/destined to dry ports. Araujo Jr. (2004) referred to this as a manifestation of a holdup problem within the context of a network industry, where a player had enough power to extort additional benefits from other players within the network. In this case, a port held the cargoes and benefitted from the dwell time of the cargoes within its premises, charged the dry port an additional fee to release cargoes, and thus placed the dry port at a competitive disadvantage, while affecting the fulfillment of the dry port's contracts with the importer. Although THC2 was found to be illegal in a recent court case that involved the port of Santos (the jury based its decision on the understanding that THC2 did not relate to any offsetting service that justified such an additional charge) (Padilha, and Ng, 2012), with the absence of a contractual relation between ports and dry ports and an effective legal framework to regulate such relationships, this condition continued to hinder the integration between ports and dry ports, and weakened the latter's logistical functions, and increased costs to importers.

Institutional reforms also affected the integration of different components along the supply chains. By increasing the number of stakeholders, planning and decision-making became more complex with a need for greater coordination. During the process, competition was created not only among private operators, but also among public agencies. For instance, the Federal Revenue regulated the dry ports and bonded areas within ports, while the Special Port Secretariat (*Secretaria Especial dos Portos*) and Ministry of Transports regulated the ports and river ports, respectively (Tecnologística, 2008). On the other hand, the institution which regulated port operations, the National Agency for Waterway Regulation (ANTAQ), was formed in 2002 – almost a decade after the ports' privatization. All these settings contributed towards further fragmentation of Brazilian logistics and supply chains. Moreover, while Law 10.233 of 2001 originally prescribed a single regulatory agency for all transport modes, later, a replacement project was sent to the Congress dividing this agency into ANTAQ and the National Agency of Land Transport (ANTT), so as to ensure that water transportation would receive due attention (Goldberg, 2009). This division, however, was criticized by the National Confederation of Industries (CNI) and the Brazilian Association of Basic Industries and Infrastructure (ABDIB) which believed that this would unnecessarily harm the articulation of projects among different transport modes, and stimulate further political controversies.

8.2   Legal framework towards institutional limbo

A number of dry ports also faced considerable uncertainties due to the lack of an adequate legal framework. Until 1995, dry ports could be established through a simple authorization process from the Federal Revenue. The situation became more complex after the Federal Constitution of 1988 was reformed in 1996, the reform declared that both ports and dry ports were state properties, and thus the federal government should explore them directly or, alternatively, through concession agreements to private firms under state regulations. Also, it stated that such concessions and exploration of public services by private companies should be given by means of public tenders (Nascimento 2005; Matayoshi 2004). However, the issue of whether dry ports constituted public or private services only became clear in 1995 with Law 8.987, which declared that any services offered by dry ports were public and thus subject to public tenders (Trade and Transport, 2004). In 1995–2003, all concessions were given by means of public tenders, effective for 25 years and extensible for a decade (Assumpção et al., 2009). Meanwhile, the existing dry ports needed to operate under emergency contracts, initially extended only 180 days. In other words, some established dry ports might face the termination of contracts by legal forces, as well as the prospect of having to go through public tenders simply to stay in business. However, after taking the case to the court, they were allowed to operate under emergency contracts without an expiry date. Meanwhile, new rulings were promulgated so as to change the legal status of dry ports which offered them new terms, definitions and expanded remits. Introduced in 1976 by Decree Law 1.455, which authorized the execution of customs clearance in secondary areas,[3] dry ports were initially called Public Bonded Warehouses (*Depósitos Alfandegados Públicos*), later renamed to Inland Custom Stations (*Estação Aduaneira do Interior*, or EADI). In fact, the term dry port (*porto seco*) was only adopted in 2002 under Decree

4543, Art. 724. In the same year, the use of dry ports for industrial operations was introduced by Normative Instruction 241/02 which suspended certain taxes for goods assembled or produced within dry ports for exports.

Responding to protracted legal disputes involving the effected dry port operators, the Provisional Measure 320 (MP320) was introduced in 2006. The objective was to end the tender requirement and to allow dry ports to operate by means of licenses issued by the Federal Revenue. This measure also expanded the scope of dry ports – to be called Logistical and Industrial Centers (*Centro Logístico e Industrial Aduaneiro*, or CLIA). However, this new measure was soon deemed unconstitutional, and thus rejected, by the Brazilian Senate – a clear indication that the initial institutional settings could not change, as the new system was deemed not in compliance with the rules of Brazil's polity and economy. Instead, it was replaced by the Provisional Measure 327 and duly transformed into a Law Project (PLS 327/2006). At the time when this study took place, PLS 327/2006 was still pending approval by the Brazilian Senate. With the tender processes currently on hold, the future development of the dry port sector in Brazil would remain rather blurred until the bill's final approval.

The institutional challenges faced by dry ports were part of a broader institutional context influenced by historical, economic and political circumstances. Protracted legislation evolved slowly in the context of an awakening democracy with a fragmented political system. As noted by Knox (2001), the design of the Brazilian institutional system complicated the passage of reform legislation. However, the comprehensive reforms achieved during the 1990s happened at a unique moment in Brazilian history. Dry port legislation had come along as a response to the new economic environment and the institutional limbo which dated back to the constitutional reform of 1988. In this regard, it should be noted that such changes in dry port laws were not unique and could not be treated as an isolated process. Indeed, the rejection of the original project reflected the limits of privatization. During an interview with *Agencia Senado* (2010), Senator Sergio Zambiasi pointed out the unacceptability to consider dry port operations – an activity with such 'national strategic importance' – to be constituted as pure economic activities, and so they must be subject to private laws and licensing. This was clearly not only a debate over the nature of public services, but a political wrestling between different public agencies and jurisdictions within the Brazilian institutional system.[4]

8.3 The bureaucratic and logistical roles of dry ports

While institutional factors created a state of inaction within the dry port sector in Brazil, due to a lack of clear and effective rules, some dry ports were established for reasons that went well beyond logistical efficiency, at times located in places that made little economic sense. With logistical inefficiency and complex trade processes, the existence of some dry ports was justified primarily by their roles in facilitating the bureaucratic functions. Here one should note that bureaucracy was necessary and could play positive roles. In this regard, the Brazilian Association of Dry Ports (ABEPRA) pointed at the advantages derived from the use of dry ports, e.g. suspension in duty payments, proximity to custom clearance process, immediate container unloading thus avoiding demurrage, permanent presence of inspection agents, lower risk of cargo loss and damage, and the possibility of partial cargo imports and exports according to company needs (ABEPRA, 2010). Dry ports could be used for a variety of custom regimes with the suspension of duties, e.g. custom entrepôt, temporary admission, certified bonded warehousing, and so on. In this regard, the suspension of duties allowed foreign firms to use these facilities as export processing bases.

The following provides an excellent example highlighting the aforementioned problem. To import or export, a Brazilian firm must be able to use the Integrated Foreign Trade System (*Sistema Integrado de Comércio Exterior*) (SISCOMEX) and be registered with the Secretariat of Trade Secretariat (SECEX). It was a system which attempted to consolidate all the required information and processes. Customs brokers were often in charge of registering and feeding the SISCOMEX system. When cargoes arrived at a port, an import declaration was registered. After the exchange rate was closed, the cargoes went through parameterization with different levels of inspection being set. A 'green' channel indicated automatic clearance. The 'yellow' channel required documental inspection only. The 'red' channel required both documental and physical inspection. Finally, the 'grey' channel required documental and physical inspection, as well as an assessment of custom values. Thus, the physical presence of a custom broker was necessary to present the required documents and to pay the necessary fees. If no problems were found, the clearance process would be registered in the SISCOMEX system by the Auditors of the Federal Revenue, and the process would be completed when the Import Confirmation was issued by the system.

On the other hand, to use a dry port, cargoes arriving at a port must be issued a Customs Transit Declaration (DTA). With this document, cargoes under customs control could be transferred from the port to the dry port, where customs inspections and the collection of duties would take place. According to the Director of the Syndicate of Customs Brokers of Minas Gerais, Frederico Pace, this was the main problem affecting dry ports within the state (Pace, 2010). The DTA removed tax collection services from one custom office to another, reducing revenue and workers in that department. For that reason, he pointed out that, in general, custom officers at the ports did not like DTA and would jump at any opportunity to block this process. In this regard, the effect of bureaucracy became inefficient and even excessive. Despite the sophistication of the SISCOMEX system, the complexity of the legislation, the large number of taxes and related processes, the exchange rate transaction requirements, and the trade restrictions and environmental regulations, combined to create a minefield for shippers where mistakes were common, leading to interruptions in logistical flows. As noted by Pace (2010), in Minas Gerais, custom brokers were structured to provide services both at ports and dry ports. If a firm enjoyed a high rate of green channel at the port with reduced bureaucracy, they became inclined to choose the ports (rather than dry ports) to clear customs. Similarly, exports were less likely to go through dry ports because they tended to face fewer bureaucratic hurdles than imports (which represented about 70 percent of cargoes cleared at dry ports). In other words, complexity in custom clearance significantly influenced the decision-making process and the bureaucratic role of Brazilian dry ports.

In Brazil, about 80 percent of imports and 85 percent of exports were given the green channel: this was among the world's highest inspection rate (the global averages were 90 percent and 95 percent, respectively). In 2008, the Federal Revenue apprehended 280 million US dollars in merchandises and applied over one billion US dollars in fines (Folha, 2009). An illustrative example can be found in the dry port of Uruguaiana in Rio Grande do Sul, where on average, about 160 trucks experienced documental or physical inspection per day. Here alone, between January 2006 and July 2008, a total of 7,467 import processes were stopped by the Federal Revenue for technical problems, for example, incomplete or incorrect documentation, incorrect entry in the SISCOMEX system, incorrect classification of cargoes, weight and volume discrepancies, fraud and under-invoicing, to name but a few. This implied nearly 1,000 cases per auditor per day with only eight auditors working in the dry port at that time (Unafisco, 2008). Unsurprisingly, according to the World Bank's annual survey on the environment of doing business in different countries and regions, in 2010, out of 183 countries and regions, Brazil was ranked 129th (World Bank, 2010).

Excessive bureaucracy, due to institutional factors, and thus complexity in the clearance process, increased demands for direct, face-to-face contacts with custom officials, amplifying the importance of the brokers to minimize custom duties, to solve problems and to speed up the process at ports and dry ports. In fact, the term broker (*despachante*) had even become part of the daily Brazilian vocabulary as a direct manifestation of excessive bureaucracy, where they played crucial roles in facilitating the interaction between state and society. In this regard, an interviewee representing a multinational freight forwarder provided the authors an illustrative example. In the US, their customs brokerage could be carried out through a paperless process from a central, inland location. On the other hand, a similar task in Brazil required the physical presence of agents at each port where they handled cargoes. Meanwhile, when negotiating rates with freight forwarders or directly with carriers, Brazilian importers generally demanded free time demurrage of 30 days, compared to an average of only five days in North America. Furthermore, ports and dry ports often suffered the impacts of strikes by officials from public agencies involved in cargo clearance, such as Anvisa, Federal Revenue, Ministry of Agriculture and Environmental Agency.[5] On average, the occurrence of strikes of any scales amounted to about 200 days per year, leading to major backlogs in ports, re-routing and very long waiting time for ships. For shippers, this represented significant insecurities in daily operations with cancelled orders, delays and high logistical costs. Under such circumstances, they were often forced to resort to informal, rent seeking practices. In this regard, an interviewee referred to the urgency fee (*taxa de urgência*) as a generic term where custom officials were often informally paid so as to prioritize, and thus speed up, the process.

According to Wu and Goh (2010), ports in a number of developing economies, such as Shanghai, Chittagong and Santos, had achieved efficiency levels surpassing those in the G7 nations. Although such ports might not have the necessary equipment endowments, they became more competitive in operational efficiency. However, in this regard, other factors such as energy supply, cost of fuel, inland multimodal connections and customs clearance might have certain impacts. When choosing a port-of-call, shipping lines might consider its inland connections. Thus, the long term viability of dry ports depended on three main attributes: (i) their geographical location where there were sufficient demands and access to high capacity corridors linked by rail; (ii) their ability to provide efficient and sophisticated value-added services; and (iii) their ability to integrate with ports and shipping lines. In areas where port development had reached an advanced stage and intense competition (cf. Notteboom and Rodrigue, 2005), hinterland access became decisive in inter-port competition. Rodrigue et al. (2010) pointed to three main criteria in the definition of dry ports: (i) the 'massification' of cargo flows; (ii) containerization; and (iii) the employment of dedicated terminals with the use of a high capacity corridor. In this context, to compete in the global market, inland production sites must be integrated with the global supply chains as much as possible, in which case the relationship between ports and dry ports is of utmost importance. While institutional deficiency fostered the bureaucratic role of dry ports, it undermined their ability to deliver efficient logistical services. Through the interviewees, the authors found that in the absence of an effective legal framework, some dry port operators felt less confident to invest in infrastructure, intermodal facilities and equipment.

Among the dry ports studied in this chapter (Table 8.1), only two used rail in a consistent manner. Such a road bias reflected the scenario in Brazil as a whole, where dry ports were connected to ports primarily by road. The dry port of Anapolis provided an exception to most other dry ports in Brazil for its intense use of rail, accounting for 87 percent of the total containerized cargoes being moved. Much of the process cleared at dry ports was linked to imported containerized cargo moved by rail, returning to ports loaded with commodities to clear customs at the ports. The problem was that most other dry ports either did not have sufficient cargoes, or lacked infrastructural conditions, for the implementation of rail. With rail being mainly used for the movement of agricultural and mineral commodities, rail operators had few incentives to implement rail services for containers. The express rail service exclusive for container transportation, provided by MRS Logistica operated on fixed routes, frequencies and transit-times. However, to use this service, a firm must contract regular weekly or monthly services with at least five wagons, or 20 TEUs, per shipment (100 TEUs per month) for at least a year. Given that most Brazilian shippers consisted of small- and medium-sized firms, in practice, few of them were able to meet this threshold.[6]

Further, the adversarial relation between ports and dry ports tended to hold cargoes at ports, affecting smooth logistical flows. As mentioned earlier, in general, interviewees from the dry ports described their relation with ports (and port terminals) as 'cold,' 'competitive' and 'problematic.' For instance, in Minas Geris, with one exception, there was little collaboration between ports and dry ports. The latter focused their efforts on obtaining the deferral of the Merchandize and Services Circulation Tax (ICMS) – a state tax that served to encourage the use of dry ports in Minas Gerais. Simultaneously, the dry port of Anapolis, located over 1,000 km away from the port of Santos, pointed to the obstructions created by ports, such as restrictive time and procedures for cargo loading, additional fees and even lobbying to avoid cargo transfers via DTA. In fact, according to various interviewees, the situation had become so problematic that it even prompted some dry ports to join forces so as to counter the aforementioned anti-competitive behaviors of ports. However, it should be noted that the institutional status quo had posed a different impact on operators owning both ports and dry ports, thus having greater control over the local supply chains and market power. Illustrative examples included the Libra Terminals in Santos (which owned the Libraport in Campinas) and Santos-Brasil (which acquired Mesquita – a transport company which also operated dry ports). In these cases, cooperation between ports and dry ports existed. Despite the existence of certain institutional obstacles, these operators continued to invest and expand. In the absence of public direction and clear rules, it seemed that these dominating groups might determine the shape of the future of dry port networks in Brazil.

Finally, the fragmented nature of supply chains in Brazil was noted in the federal government's inability to effectively implement a recently passed law – the Multimodal Transport Operator (*Operadora de Transporte Multimodal*) (OTM)[7] – which required shippers to use at least two or more transport modes from origin to destination under one single operator and transport bill. When this study took place, there were about 800 OTMs in Brazil. According to Keedi (2010), this issue was in the hands of federal government, with neither the Central Bank of Brazil, nor the Federal Revenue, recognizing OTM. Thus, it still faced considerable uncertainties related to the insurance policies covering the entire logistical process with so many stakeholders being involved. According to the interviewees from the Special Port Secretariat, the incentives to block the effective implementation of OTM were mainly operational considerations, notably the determination of responsibilities for cargo damages, and the lack of standard practices concerning the application of custom duties across state jurisdictions.

8.4   The case of Sao Paulo

To enhance the understanding on the effects of institutions, this section discusses and analyzes the spatial evolution of ports and dry ports in the state of Sao Paulo, and focuses on how it has affected the evolution of port and dry port configurations in this state in the past decade. The state of Sao Paulo hosts one of the world's largest metropolitan areas and the port of Santos is Brazil's leading port. Unsurprisingly, a substantial portion of the Brazilian dry ports were established here (Figure 8.1). In this case study, two main questions were investigated: (i) how did the development of the port of Santos affect the spatial evolutionary pattern of dry ports in the state of Sao Paulo?, and (ii) Why were dry ports in the state of Sao Paulo not been able to develop in line with Brazil's economic growth and the patterns of port development? This analysis can offer valuable insight to scholars, policymakers and industrial practitioners, in particular regarding the development of dry ports in other emerging cities and developing economies where dry ports were rapidly developing. As mentioned in Chapter 7, only few studies on the development of dry ports in emerging economies exist, including Latin America.

8.4.1   The evolution of the port of Santos

The port of Santos was founded in 1892, initially focused on the export of coffee brought by rail from the interior part of the state of Sao Paulo. During the first half of the last century, cargoes were predominantly moved by rail. The situation gradually changed since 1947 when *Rodovia Anchieta*, a road connecting Santos and Sao Paulo, was constructed. This marked the beginning of a road-bias policy at the expense of rail and cabotage,[8] followed by the construction of *Rodovia Imigrantes* in 1974. Over the years, these roads have been further expanded so as to meet the rising demands from the port of Santos and the increasing traffic from a growing metropolitan area, i.e. the city of Sao Paulo. The roads remained the two main arteries linking the port and the city, as well other parts of the state of Sao Paulo. After the 1950s, port activities increased due to increasing industrialization with the construction of refineries and automotive industries. Transport projects and investments led to an extensive road network, particularly in the metropolitan area, and thus the fleet of cars and traffic also increased simultaneously. Meanwhile, cabotage sharply declined and paralyzed the close connections between Santos and other Brazilian ports.

The influence of containerization did not take long to affect Brazilian ports, including Santos, but the pace of development was slow. Although Santos started to receive containerized cargoes in the mid-1960s, the first container terminal was inaugurated only in 1981 controlled by *Empresa de Portos do Brasil SA* (*Portobras*). According to Pimentel (2004), with the transition in port administration due to the termination of the concession period from *Companhia das Docas de Santos* (CDS) to the control by the newly created *Companhia das Docas do Estado de Sao Paulo*, institutional transformation was coupled with philosophical changes in response to domestic forces and international practices. At the same time, congestion problems led to further expansion. As the Brazilian economy became more sophisticated, there was a gradual increase in container throughputs in the port of Santos.

Concentration continued with improvements in technology, container handling equipment and the expansion of dedicated areas. However, the integration process of ports into the supply chains was very slow. A main reason was because inland transportation was almost entirely undertaken by trucks, with rail being used predominantly for the movement of agricultural and mineral commodities.

Until the early 1990s, the development of the port of Santos towards a higher level of sophistication and integration was seriously hampered by strict central and public control, with port activities concentrated in the hands of Portobras and administered by dock companies, though the port often suffered a shortage of financial resources for capacity investments (Galvão et al., 2013). The situation started to change in 1993 with the dissolution of Portobras and the promulgation of the Port Modernization Act (8630), with the first lease agreements under a new model being implemented in 1995. This act paved the way for private participation in port terminal operation, created incentives for new investments in infrastructure and equipment, and reforms to demolish the monopoly of dockworker unions. From 1996 to 1999, reforms were also undertaken to break up the national rail network, which consisted of 28,445 km, into 11 parts operated by private firms through a number of concession agreements. Despite such improvements, rail was still rarely being used for container transportation with few connections to dry ports.

As mentioned earlier, the centralized nature of port governance, restrictions to private investments, the decline of cabotage and the impediments to foreign investments in these areas blocked the evolution of feeder ports or any eventual challenges from the periphery. The limitation of cargo flows to the two main highways (*Anchieta* and *Imigrantes*) over the past decades created an explosive situation with no immediate solutions. Although the port of Santos grew to become the largest container port in Latin America, it remained virtually unchallenged within the state of Sao Paulo. It reached capacity limits with significant negative implications to the city and its metropolitan area. Infrastructure limitations, together with institutional barriers and inappropriate policies, ensured continued concentration within the port of Santos. Without any alternative feeder ports to absorb at least some excessive demands, de-concentration manifested itself with the emergence of dry ports. Also, with the absence of high capacity rail corridors, instead of the emergence of large scale inland hubs located at intermediate positions with the ability to consolidate freight from more distant areas, as witnessed in other developed economies (as discussed in Chapter 7), the dry port configuration in the state of Sao Paulo consisted of a relatively high number of small facilities serving a rather restricted hinterland.

8.4.2   Factors affecting the development of dry ports in Sao Paulo

Institutional system: A number of institutional deficiencies affecting the ability of dry ports in Sao Paulo to concentrate cargoes and integrate into the multimodal supply chain can be identified. Most dry ports were established after the 1990s when the Brazilian government's experiment with economic stabilization and trade liberalization, revealed its inadequacy on transport and logistics infrastructure. Despite significant modernization of the port system, which various interviewees agreed had occurred, it was clear that dry ports were often treated as being of secondary importance and were operated under an environment of institutional uncertainties. In this regard, three major problems could be recalled. First, as mentioned earlier, the regulation of the sector by different institutional agents contributed to fragmented policies and conflicts of interest. Second, the relationship between ports and dry ports was not adequately supported by legislation, thus the legislation failed to provide clear rules and regulations regarding cargo transit between these two types of ports. Also, existing legislation failed to create incentives for cooperation, thus triggering ports and dry ports to compete with each other for similar logistical and value-added activities, notably custom clearance and storage revenue. In this situation, congested ports were often used as storage facilities, with ports (and terminals) often creating difficulties, or even imposing surcharges, this situation undermined the effective operation of dry ports. For instance, one of the dry ports located within the port of Santos (operated by Integral) recently ended operations, alleging the need to increase capacity at other terminals. The firm referred to high operational costs, including the aforementioned THC2 charged by port terminals for cargo transiting to dry ports. Finally, a number of dry ports faced considerable uncertainties regarding their legal status. Until the early 1990s, dry ports could be established within Brazil by means of licenses given by the Federal Revenue. However, their status became blurred after the Federal Constitution of 1988 designated ports as state assets which might be operated directly by the federal government, or by the private sector through concession agreements awarded through public tender (see Section 8.2). Disputes over the status of dry ports persisted until 1995, when Law 8.987 clarified dry ports as public facilities, subject to tender so as to operate legally. This decision triggered several legal challenges as dry ports with effective contracts saw their operations facing serious jeopardy by legal forces. To make the issue even more complex, there were four dry ports in Brazil established during the short effect of the aforementioned Provisional Measure 320 (2006), which expanded the scope of dry ports, to be called CLIA, by ending the tender requirement and re-introducing the license system. However, as mentioned earlier, the Provisional Measure 320 (2006) was later deemed unconstitutional by the Brazilian Senate, and thus was in effect for only four months; the sector's development was still in limbo when this study took place.

Intermodalism and integration: There were few dry ports in Sao Paulo which had rail connections, although some attempted to initiate rail services with certain degrees of success. For instance, Jundiaí (37,500 m^2) had a 160 km rail line built in partnership with MRS Logistica connecting its dry port with the port of Santos. A main user of this service was *Grupo Votorantim* carrying about 800 TEUs per month to Santos Brasil Terminal (Guaruja) at Santos. Also, the dry port of Libraport in Campinas[9] invested four million *real* in constructing a 2.6 km rail line linking it to the main rail line. Meanwhile, another initiative was the establishment of a rail service by ITRI – *Rodoferroviae Serviços Ltda.*[10] between the port of Santos and the dry port of Cragea in Suzano. Despite these efforts, other dry ports were less successful in developing rail services, and thus multimodal transportation. For instance, although Bauru had a rail service (the Bauru-Corumba railway) that was extensively used for the export of steel products for Arcelor-Mittal between Bauru and Bolivia, its dry port did not provide any support to rail services, partly due to the high threshold required by operators as mentioned earlier.

Indeed, the ability of dry ports to develop agglomeration largely depended on to what extent their facilities could integrate different transport modes, especially road and rail, and access to high capacity rail corridors. As mentioned earlier, in the state of Sao Paulo, the transportation of containerized cargoes by rail was still incipient, with most of the networks being only used for mineral and agricultural products. Until the mid-2000s, only 1.9 percent of the containers arriving at Brazilian ports and 1.6 percent of containers leaving from Brazilian ports were moved by rail (Hijjar and Alexim, 2006). Although this rate had doubled when compared to the time period 2001–2005, it was still relatively small, reaching just over 100,000 TEUs in 2005, compared to 11 million in Class I railways in the same year in, for example, the US. In fact, privatization and improvements to the rail sector had largely benefitted mineral and agricultural products rather than containers, with iron ore representing over 74 percent of total movement in 2009. According to the Director-General of ANTT, such concentration had accelerated over the past years. Meanwhile, despite the increase in container rail transportation in recent years, they only represented 4 percent of the total freight movement (*O Estado de Sao Paulo*, 2010). The largely imbalanced modal distribution in Brazil and the state of Sao Paulo can be found in Table 8.2. As pointed out by various interviewees, rail operators preferred to carry commodities with large, regular volumes over long distances, but most shippers of containerized cargoes in the state of Sao Paulo, like most other Brazilian states, consisted of small- and medium-sized firms, and most of them found it difficult, if not impossible, to meet this threshold alone. Hence, in the port of Santos, most containers were still moved by trucks, with rail only responsible for about 18 percent. Despite some improvements thanks to privatization, rail services continued to tackle

Table 8.2 The distribution of transport modes in Brazil and the state of Sao Paulo, 2006

Transport Mode	Brazil*	State of Sao Paulo**
Road	59.0%	93.1%
Rail	24.0%	5.3%
Water	13.0%	0.5%
Air	0.3%	0.3%
Pipeline	3.7%	0.8%

Sources: *Ministry of Transport, Federal Government of Brazil; ** State Secretary of Transport, Government of the State of Sao Paulo.

considerable hurdles related to the conditions of the network, insufficient intermodal facilities and gauge differences. Also, as the Brazilian rail concession system did not separate operation and infrastructure, operators were given exclusive rights to operate in their respective segments, and this posed significant challenges when cargoes transited through rail lines under different control (Lacerda, 2005).

In the state of Sao Paulo, rail service was further complicated by the sharing of rail lines between cargoes and passengers; this situation discouraged operators from increasing capacity. Also, there were problems of accessibility to the port of Santos which led to the creation of Portofer in 2000, with concession for 25 years, to improve rail access, operations, facilities and equipment. Although Portofer made some improvements to reduce the staying time of wagons at the port (from 96 to 32 hours), and to reduce the number of port-related accidents, a lot remained to be done to reduce congestion and to improve accessibility.

Urban congestion and limitations: Some dry ports were affected by (and also contributed to) urban congestion, this limited their scope of agglomeration and integration even further. The 25 dry ports established within the state of Sao Paulo were operated by more than 15 different firms, with 20 located within 100 km from the city of Sao Paulo, and nine of them within the metropolitan area.[11] They typically provided facilities for storage and cargo handling, refrigerated areas, plugs for reefers, scales and lifting equipment, patios for truck maneuver and container storage. However, most of them were small in size when compared to many of those in other developed economies (Table 8.3).

Table 8.3   A comparison of the average sizes of dry ports between selected countries, 2010

Country	Average size in acres
Brazil	15
Denmark	257
Germany	200
Italy	766
Spain	133
UK	393
US	3088

Source: Adapted from various industrial sources.

Largely due to such small sizes, many dry ports in the state of Sao Paulo failed to offer cost advantages that might attract shippers from further afield which contributed to the consolidation of cargoes. Apart from institutional deficiencies and the absence of high capacity rail corridors, they were further constrained by urban congestion. Indeed, transport infrastructure within the state of Sao Paulo was concentrated in its metropolitan area, and was largely created to serve the city of Sao Paulo – an economic focal point of Brazil, and a critical communication point with the international market – and evolved as a reaction to the city's expansion. As the city of Sao Paulo grew, its impacts reached beyond sustainable limits, thus prompting the emergence of nearby regions.[12] An illustration of the spatial setting of dry ports in the state of Sao Paulo can be found in Figure 8.2. One can compare it with the typical spatial setting of dry ports/inland terminals in advanced economies, for example, the US (Figure 8.3).

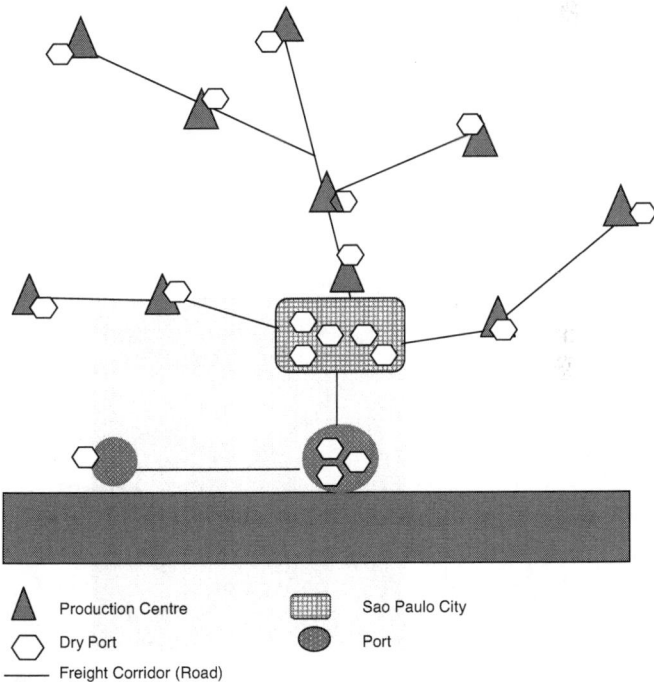

Figure 8.2 A diagram illustrating the spatial settings of dry ports in the state of Sao Paulo, Brazil

Source: Authors.

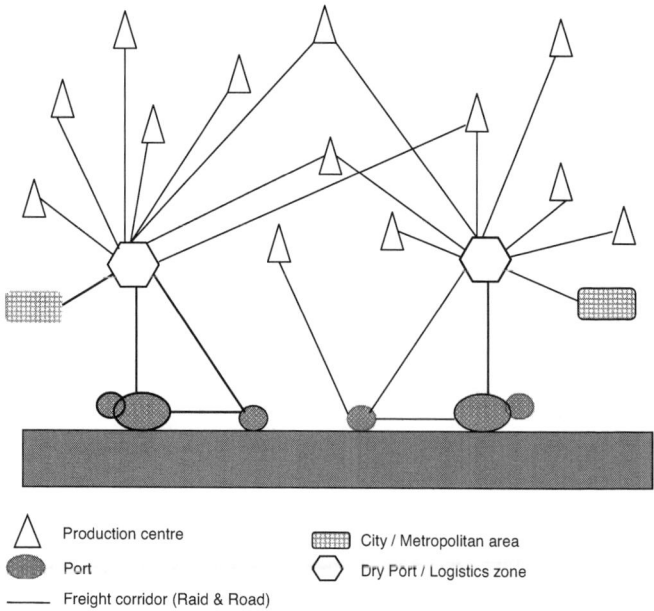

Figure 8.3   A diagram illustrating the typical spatial settings of dry ports in the US

Source: Authors.

For instance, the three dry ports located near the city core – CNAGA, Columbia and Embragen, with areas of 64,000, 18,000 and 22,000 m², respectively – were all located within the industrial regions with very little room for expansion, while some other dry ports were located next to the so-called 'shanty towns.'[13] Under such conditions, it is not surprising that the creation, or particular concentration, of logistical services or value-added activities by dry ports in the state of Sao Paulo have been found wanting. Although they served the industries and production plants surrounding them, serious traffic congestion strictly limited their catchment areas. Where these problems were compounded, some dry ports were forced to close down, as was the case of the dry port of Guarulhos (20,000 m²) controlled by Deicmar (a customs broker and warehouse operator which also operated a dry port in Santos with an area of 65,000 m²). After a decade of operation, in 2010, it announced the cessation of the dry port activities in Guarulhos.

8.5 Discussion and conclusion

The situation in Brazil provided another illustrative showcase that ports and dry ports were strongly affected by institutional settings and reforms. While institutional settings and reforms established the foundations for necessary bureaucratic processes and logistical modernization, simultaneously, they were the causal factors of a multitude of problems which affected port and dry port development. In this regard, they had a critical, and detrimental, effect on the integration between ports and dry ports, and further posed negative consequences to the overall economic development and the competitiveness of Brazilian merchandises in the international market. Also, the existing regulatory framework provided disincentives for ports and dry ports to cooperate. Where dry ports could possibly increase the efficiency of ports, their adversarial relationships fragmented Brazil's rather embryonic supply chains, with ports being placed in a dominating position and serving as storage facilities, rather than the catalytic and logistical functions that contemporary logistical hubs should possess (cf. Nam and Song, 2011). The involvement of multiple institutional agencies and regulatory bodies in the governance of ports, dry ports and other logistical sectors made supply chain integration even more difficult. However, according to the interviewees from the Special Port Secretariat, this should perhaps be interpreted from the historical perspective, where the segmentation of regulation of different port types was an anachronism that paralyzed efficiency due to fragmented and disconnected planning, and the lack of any real historical coordination among different public administrative entities. The notion that ports (including dry ports), logistics and supply chains should no longer act as cost generators, but as sources of productivity and of value-added to be exploited (cf. Galvão et al., 2013) through the implementation of logistical functions, was difficult to take off in Brazil.

Excessive bureaucracy within ports, the devastating impact of labor strikes and the complexity of clearance processes with multiple taxes, bonded regimes and trade restrictions created an uncertain environment where unpredictability led to high inventory levels and demands for extended free time for expected cargoes. Moreover, it contributed to urban congestion and further affected the smooth movement of goods between ports and dry ports. As mentioned in

Chapter 1, this altered the behaviors of shippers who placed more emphasis on informal practices and interpersonal relationships so as to overcome excessive bureaucracy. This was something very similar to what had happened in India (see Chapter 7). In Brazil, an illustrative manifestation of such bureaucratic obstacles was the increasingly prominent roles of custom brokers who, in practice, acted as the ultimate decision-makers in arbitrating the locations for custom clearance and other logistical activities. The impacts of institutional factors on port and dry port evolution and development are summarized in Table 8.4.

A more detailed analysis on the state of Sao Paulo suggested that the spatial pattern of port development commonly identified in the existing literature, mainly based on the experiences from the developed world, might vary significantly in developing economies where institutional barriers were pervasive. For instance, intermodality was poorly developed and there was almost no use of rail for container transportation. Also, there were no challenges from alternative (secondary) ports to the port of Santos with the decline of cabotage and institutional restrictions on the emergence of alternative ports in the past decades. Without any relief channels for port concentration in the form of feeder ports, de-concentration manifested inland and was transmitted through already overburdened roads. The consequence was the establishment of many small-sized dry ports with very limited catchment areas, thus unable to benefit from, or to instigate, agglomeration. These dry ports remained static due to institutional gridlocks, with few (if any) of them able to emerge and became the focal points for the establishment of integrated logistical systems and efficient supply chains.

Apart from Brazil and India, such a phenomenon also existed in Indochina (see Do et al., 2011). Hence, this chapter provides further evidence that institutional factors had acted as a significant *environment heterogeneity* which distorted outcomes. The process of port concentration in Brazil took place, despite the marginal role that the port of Santos played in international shipping, the relative absence of dominating (multinational) terminal operators in Brazilian ports, or shipping lines, in the provision of inland transportation services. This was a process largely driven by domestic institutional forces and with the main magnet and driver of port

Table 8.4 A summary of the impacts of institutional factors on the evolution of ports and dry ports

Institutional issues	Port/dry port development	Impacts to logistical role	Impacts to bureaucratic role
– Legislative uncertainties and inactions concerning laws that regulated ports/dry ports	– Difficult to install new port/dry ports, while existing ones are unlikely to invest, and some may be forced to shut down	– Hinders new investments in infrastructure, intermodal facilities and equipment	– Preserves long-standing excessive bureaucracy within ports/dry ports
– Competing government agencies in charge of different stages in the supply chain	– Absence of integrated planning which leads to an inefficient network of ports/dry ports	– Fragmented and inefficient supply chains, with each stakeholders possessing different (or even conflicting) objectives and targets	– Preserves long-standing excessive bureaucracy within ports/dry ports
– Inadequate laws to enhance cooperation between ports/dry ports	– Dry ports exist in isolation disconnected from port terminals (except those under the same ownership)	– Adversarial relationship between ports/dry ports – Use of port terminals as storage facilities, thus interrupting the efficient flow of cargoes	– Additional difficulty with cargo transfer; bureaucracy between ports/dry ports
– Excessive taxation, protectionism, complex trade regulation, strikes	– Promotes the development of small scale ports/dry ports (sometimes at places that make little economic and/or logistical sense)	– Hinders the efficient flow of cargoes – Hinders the use of rail in favor of trucks – Hinders the creation of intermodal facilities – Increases time of loading and unloading of cargoes at ports/dry ports – Congestion – Hinders cargo transfer – Increases the demand for free time demurrage at ports/dry ports	– Establishes a necessity for face-to-face contact with custom officials, which increases informality and risks.

Source: Authors.

concentration and development being the agglomeration forces of the metropolitan area itself.

With institutional factors being the main determinants of transport infrastructure bottlenecks, some key policy recommendations could be well understood in Brazil but difficult to implement due to entrenched vested interests and political forces. Policies to develop cabotage and rail transportation should remove the current imbalance in favor of road transport while delivering significant environmental gains. There is an urgent need for integrated planning, clear rules to encourage investment and the elimination of institutional gridlock. Also, legislation should encourage cooperation between ports and dry ports, and discourage shippers from using ports as storage facilities. Adequate planning should also establish the groundwork for the deployment of strategically positioned dry ports connected by rail creating an environment conducive to economies of agglomeration. The *PDDT Vivo 2000/2020*, a plan conducted by the Secretary of Transports of the state of Sao Paulo and by Dersa SA could go a long way towards meeting this goal. The study, which identified 70 bottlenecks along the transport networks, was discussed with public and private stakeholders, and suggested the creation of inland logistical hubs around the metropolitan area, and at other major intermediate inland locations, served by high capacity rail and road corridors. It recommended actions, such as a construction of a road ring with 170 km, linking ten highways around the urban core of the city of Sao Paulo and a rail ring around the metropolitan area, the implementation of an express freight rail service, incentives to cabotage with integration of the ports of Santos and Sao Sebastiao and transfer of the administration of the port of Santos from the federal government to the state and municipal governments (Braga, 2008). Meanwhile, some firms undertook vertical integration with key logistics stakeholders so as to control supply chains and overcome institutional barriers, with the aforementioned Libra Group (and its dry port, Libraport Campinas) being an illustrative example. When this study took place, this group, which also dominated terminal operations in the port of Santos, planned to invest 440 million *real* (about 193 million US dollars) to expand its container terminal area by doubling the current capacity from 900 to 1,800 TEUs.

The above integration strategies might help to overcome problems caused by institutional fragmentation and the competitive relationships between ports and dry ports, although questions remain as to the alignment of such initiatives with long term national and state transport infrastructure planning and policies. The successful implementation of the above strategies in Brazil, and indeed many other developing economies, would require innovative, effective solutions to address the aforementioned institutional hurdles. After studying many cases in previous chapters, the concluding chapter will summarize the research findings and ideas in this book, as well as suggest the appropriate way forward for ports, logistics and supply chains.

9
Port-Focal Logistics:
The Ideal for Future Global Supply Chains?

The analysis and findings from previous chapters clearly illustrate that global supply chains have been revolutionized in the past decades, and that ports have played, and will continue to play, much more decisive roles in the well-being of logistics and supply chains. Based on this understanding, the chapter will summarize the major findings of the text to provide sound justification for the main idea that this text highlights for readers: port-focal logistics as the appropriate development direction for future global supply chains. In addition, the chapter will also emphasize the potential contributions of this book to future research regarding the development of ports, logistics and supply chains; in particular, this book will serve as the perfect platform for further research on this topic.

9.1 The need for port-focal logistics

This book has illustrated the factors that have contributed to the transformation of ports from simple sea-land interfaces to integrated components into global logistics and supply chains. Many ports have improved their connections with hinterlands, as well as inland transport and logistics infrastructures. Simultaneously, within the port arenas, rather than just cargo loading and unloading, many started to provide more diversified, and value-added services and targeted more users from different industrial sectors instead of just shipping lines. Indeed, a number of major ports around the world have undertaken neoliberal reforms so as to enhance efficiency, to share financial commitments from the public sector, as well as to respond better to the demands of users. This has opened the door to ports to become more integrated into logistics and supply chains, while they are also offered a good opportunity to transform themselves into logistical hubs with a more hybrid community. However, several challenging questions have yet to be satisfactorily addressed: should ports encourage the concentration of logistical and value added activities around themselves; or should they strengthen their bonds with inland market, transport and logistics infrastructures, especially dry ports, so as to sustain their attractiveness to users and, ultimately, to enhance their competitiveness? Even within the reform process, given the on the ground realities within different countries and regions, notably diversity in political traditions and institutional systems, will a similar solution, when applied to different cases, lead to the same outcome? With the potential pitfalls of increasing governance complexity, is there an international 'best practice' that can fit in the new circumstance, or does the choice, and appropriateness, of the reform approach actually depend on the regional and local circumstances?

The massive wave of globalization ensures that logistics and supply chains have become more multinational, and have clearly evolved from being 'firm' focal to 'port' focal. There have been various discussions on how such a port-integrated logistical system should evolve, such discussions range from the penetration of ports to inland regions (and the establishment of dry ports/inland terminals) to the development of port-centric logistics where logistical and value added services should remain within the port's arena (which also involve the increasing collaboration between proximate ports to form regional port clusters so as to strengthen their capacities in providing such services). However, in this book, the authors argue that while these arguments possess some merit, they are not adequate in sustaining the competitiveness, and indeed survival, of ports in the long term; port development must be multi-directional. Like Hong Kong and the PRD (see Chapter 4), it is clear that a port-focal logistical system should be established, with ports being the focal points in the development of logistics and supply chains.

Moreover, in many cases, when assessing the performance of ports, port-integrated logistical systems and supply chains, it is found that *environment heterogeneity* is a key factor, notably geographical, demographic and institutional characteristics, which have largely been excluded from existing assessment models. As illustrated in Chapter 5, the similarities and differences between Singapore and Hong Kong serve as excellent examples. Once again, it should be noted that *environment heterogeneity* is identified in this book as an attribute of inter-stage transactions, whereas they hold the same degree of *asset specificity* as defined in transaction cost economics theory. Recognizing this, the authors further investigate the effects of institutions as a key attribute of *environment heterogeneity*, and have conducted case studies on port reform and governance, the development of ports, dry ports/inland terminals, and their integration into logistics and supply chains, in both developed and developing economies. The studied cases confirm that institutions have significantly constrained, and retracted, actions, though institutional constraints are not necessarily the sole cause of outcomes. For instance, in South Korea and the Netherlands, exogenous factors like technological and economic progress exerted substantial pressure for ports to transform respective port governance systems, as illustrated by the corporatization of the port authorities for the ports of Busan and Rotterdam into PACs. Implementing adjustments within diversified institutional frameworks, the policymakers in both ports have moved forward towards a similar direction in management and governance structure, and new systems aimed at financial autonomy, diluting the so-called 'public' image and the inclusion of previously peripheral players in forming a network of self-governing actors which all participated in port governance, and the future direction of port development. However, the two ports also provide convincing exposition on how the diversified institutional systems can, and will, divert a generic approach in solving a globally proximate problem to different development directions. On the surface, both the ports of Busan and Rotterdam seem to have moved towards a converging governance system but, in reality, these ports found it very difficult to divert from their established institutional systems, with the South Korean port retaining a state-developmental system while theDutch port remained entrepreneurial. Indeed, at the time that these studies took place, both ports were still 'locked-in' (Pierson, 1993) by their own established institutional systems.

Moreover, the case studies from India and Brazil illustrate the pivotal role of institutions in affecting the effective development of port-included integrated transportation and logistical systems. In this regard, regional-specific background, culture and local interests can lead to different interpretation of the same concept. Indeed, both Indian and Brazilian ports and dry ports are strongly affected by institutional reforms at different levels. While institutional reforms establish the foundations for necessary bureaucratic processes and logistical modernization, established institutions also serve as the core *environment heterogeneity* triggering a multitude of problems affecting the development of ports, logistics and supply chains in these countries. In turn, these problems create a critical, and detrimental, effect to the integration between logistical nodes, and pose negative consequences to the overall economic development and competitiveness of domestic products in the international market. In some cases, these even create disincentives for logistical hubs to collaborate and think from the 'chain' perspective. In turn, this dissipates the logistical, and value addition, functions of these hubs, while strengthening their bureaucratic functions. In Brazil, where dry ports can possibly increase the efficiency of ports, their adversarial relationships have further fragmented the country's rather embryonic supply chains, with ports being placed in dominating positions and serving as storage facilities, rather than facilitating the logistical functions that modern logistical hubs are expected to perform.

More importantly, the Indian and Brazilian experiences provide some early indications that in countries and regions where *environment heterogeneity* (like institutional factors, see more in Chapters 3 and 6) are strong, the bullwhip effects may trigger the development of certain strategic countermeasures by stakeholders which may cause the bullwhip to 'repulse' and thus hurt other players in the process, either by design or accident. The bullwhip effect refers to the phenomena that the fluctuation of production flows, thus volatility and risks, magnifies when moving from the market end upstream towards the supply end of a supply chain. This usually results in, production inefficiency and excessive inventories. As mentioned in Chapter 8, the implementation of THC2 by Brazilian ports on (inland) dry ports so as to encourage shippers to undertake custom clearance and other logistical activities within the port areas (rather than in the dry ports) serves as an illustrative example. Facing such a challenge, as mentioned in the same chapter, several dry ports started to form strategic partnerships, and joined forces horizontally, as countermeasures. A similar situation also took place in Southern

India. To protect the state's local port (Tuticorin) against the port of Cochin (located in the neighboring state of Kerela), the state government of Tamilnadu aimed to discourage shippers based in Tirupur (see Chapter 7) to use the port of Cochin for export through various measures, notably state subsidies and 'credit facilities' (cf. Ng and Gujar, 2009).[1] This was despite the fact that, in pure monetary terms, the cost of moving goods towards the port of Cochin was more than 50 percent lower than that of the port of Tuticorin (Ng and Gujar, 2009). Under such phenomena in both cases, the behaviors of shippers were inevitably affected, and in turn placed even more emphasis on informal relationships, face-to-face contacts, and certain rent seeking mechanisms.

Hence, when situations similar to the above cases happen, supply chains may not only become inefficient and fragmented, but stakeholders may actually initiate a self-destructive process which accelerates the meltdown of supply chains, or destroys it altogether. When such 'competitive strategic partnerships' start to take place, instead of relieving the bullwhip effects, they may add further problems to it. As illustrated by the Indian case as mentioned earlier, it is possible that stakeholders may sometimes even want the supply chains to remain inefficient and fragmented so as to facilitate their ability to achieve certain alternative interests, both political and/or commercial. Under such a phenomenon, the 'chain' awareness among stakeholders will be blurred, thus dashing any further hopes of establishing the 'chain' culture among the logistical and supply chain communities. In fact, this also partly explains the inadequacy of existing theories on port evolution and development, like the 'outside-in' strategy in establishing dry ports/inland terminals by seaports (Wilmsmeier et al., 2011) and port-centric logistics (see Chapters 1 and 3), as they are likely to find it difficult to explain such 'repulsive' bullwhip effects convincingly, especially under situations where the effect of *environment heterogeneity* is significant.

The above proposition, of course, is subject to further research and validation. Nevertheless, the experiences from two major developing economies of the world have certainly highlighted the need to enhance the quality of coordination between logistical and supply chain stakeholders, as well as more positive communication between supply chain operators and regional planners. In this regard, who should (and can) take the initiative and interactive roles so as to effectively tackle this challenge will be subject to debates and discussion. The authors believe that ports (including seaports, airports, dry ports/inland terminals and other logistical hubs), because they are nodal points along transportation, logistics and supply chains (and centers of communications between different regions), are one of the, if not the, most competent candidates to undertake this role, and act as the catalyst spreading the 'chain' ideology among supply chain stakeholders. Such an observation may also reflect the deficiency of the 'firm' focal logistics and supply chains (see Chapter 3), which strengthens the proposition that the bullwhip effects on 'port' focal logistics and supply chains may not only be more significant, but also more diversified. For instance, in firm focal supply chains, researchers have found that integration (and thus integrative strategies) along a firm focal supply chain can be an effective solution to reduce bullwhip effects. Yet how to integrate port-focal supply chains so as to relieve the more significant, but also more diversified, bullwhip effects poses a new and wide open question. This is a serious research topic with huge potential and will pose a highly significant, and interesting, direction for future research, e.g. characteristics, measures, impacts, empirical verification, solution methods. There is little doubt that this book will offer a perfect platform for researchers to do so in the near future.

Moreover, all the studied cases warn against the attempt of imposing generic solutions to different countries and regions even when addressing a highly similar challenge. Simultaneously, they offer strong indications that the institutional system is a key variable in defining the transformation, evolution and successful establishment of the port-focal logistics. Based on this evidence, port-focal logistics can thus be understood as an integrated logistical system supported by the effective operation and governance of interrelated logistical hubs (ports). In this regard, the effective operation and governance of ports would largely depend on how the *environment heterogeneity* is manifested by policymakers and industrial stakeholders when developing their strategies and tactics. Here it is important to note that there must be ways to ensure that all the 'hubs' within the system will evolve from a 'firm' perspective to the 'chain' perspective in order to play the role of the honest brokers liaising between logistical and supply chain stakeholders, both locally and globally. Such an evolvement is especially important, given that the diversification of the internal structures of different ports strongly hinted that they might pose different impacts on the integration process of ports into supply chains. This will be especially prominent in developing economies where, as mentioned earlier, the influences of institutions are often strong.

An important inspiration from this project, as reflected in this book, is the recognition that many previous research works on ports (which included seaports, airports, dry ports/inland terminals and other logistical centers) usually viewed the development of hubs as an 'extension' of other hubs (e.g. the establishment of inland terminals in North America and Western Europe as an infiltration strategy of seaports), or view ports as independent, physical units which sought to sustain their own competitiveness. To put it philosophically, rather than just a set of inanimate facilities and infrastructures, ports nowadays have evolved into 'organic beings,' where their signaling and self-sustaining processes can, and will, pose significant implications on the health of other ports, logistics, supply chains, and ultimately, the economic well-being of the world, and vice versa.

Hence, through the proposed port-focal logistics concept, the authors strongly urge scholars, policymakers and industrial practitioners to put their emphasis on the fundamental 'lives' of ports, and how they should interact with the 'lives' of other stakeholders so as to intensify the vivacity of logistics and supply chains as the phenomenal 'organic existences' of the world economy. There is a necessity to thoroughly understand the interactive dynamics between different ports, other logistical components and regional development, and an urgent need to re-interpret the epistemology of ports. The situation will only become even more complex and uncertain in the future, given the rise, and significance, of new, challenging issues in the development of ports, logistics and supply chains around the world, for example, issues involving safety and security, climate change and sustainability (Illustration 9.1),[2] the increase in inter-regional trade, the blurring of the North-South divide and, of course, the transformation of economic and business environments permanently re-shaped by recent global and regional financial crises which took place in the past several years. Hence, achieving the ideal for efficient future global supply chains will ensure a highly challenging environment for scholars, policymakers and industrial practitioners

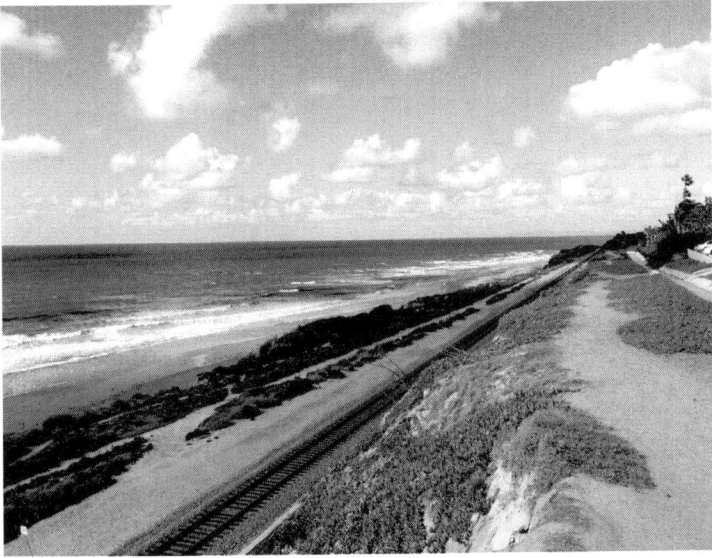

Illustration 9.1 Transport and logistical facilities, such as rail tracks constructed along shorelines, can be vulnerable to the risks posed by climate change

Source: Authors, taken in Del Mar, California, USA (2012).

to adapt and overcome in forthcoming years, and the transformation from 'firm' to 'port' focal logistics will be one of the most effective means to do so.

9.2 Epilogue

Before ending here, once again, the authors would like to highlight the contributions and added values of this book to ports, logistics and supply chains, notably how it stands out from most existing publications on a similar topic. This book stresses the pivotal roles of ports in the development of efficient global logistics and supply chains. However, despite their critical role serving as the nodes for where different logistics and supply chain stakeholders interact, ports are one of the most overlooked components. Moreover, understanding that ports, in nearly all cases, are geographically embedded within particular locations, their operation, management and governance will, as already discussed, inevitably be affected by *environment heterogeneity*, notably institutional factors. This is something that many existing publications, and indeed previous research, have overlooked.

Last but not least, this book is one of the very few exceptions which pays substantial attention to the experiences of developing economies from different continents. Apart from Western, advanced economies (e.g. the Netherlands), substantial parts of the discussions (and examples) focused on the experiences from emerging/developing economies in Asia and Latin America (e.g. China, India and Brazil). By doing so, it provides a very solid basis to offer a new theoretical approach (like the inspiration on the nature of the bullwhip effects, see the earlier section), and niche, to study, research and understand the governance of ports, logistics and global supply chains. It is a 'just-in-time' publication offering highly interesting and innovative insight to the hot debate on whether there is a developmental need, or pure coincidence, that developing economies have, in general, uneven distribution of economic masses. As a consequence, it fills a significant knowledge gap on a serious challenge that the transportation, logistics and supply chain sectors need to overcome, while simultaneously opens up a paradigm-shifting ideology for theoretical research on this important topic. The small, pioneering, step taken by this book will serve as a large and significant step for the well-being of the next generations.

Notes

1 Introduction

1. For an explanation of the impacts of the bullwhip effect on the management and integration of supply chains, see Forrester (1961) and Lee et al. (1997).

2 Contemporary Development of Shipping and Impacts on Ports

1. The introduction of TEU was inspired by the British journalist R.F. Gibney in 1969 as a solution to the challenge of recording shipbuilding and freights with different sizes and dimensions of containers used by various shipping lines in the UK. His contribution to the shipping industry should not be underestimated because, without this innovation, the speed of containerization would likely have been paralyzed by competing standards between different countries, regions and shipping firms during its embryonic stage. For further details, see Slack (1998).

2. Here one should note that ship sizes had actually been increasing since the introduction of steamships. For instance, Kaukiainen (2012) argued that, in the long run, the most central feature of shipping since the nineteenth century had been the growth of ship sizes. Nevertheless, containerization substantially accelerated this trend.

3. Throughout the past decades, there was always some concern raised by scholars over the low capital return and excessive capacity in the container shipping industry, but there is yet no fully satisfactory explanation of why there have always been continued investments pumped into this sector. For further details, see Lau et al. (2013) on the trends in container shipping research in the past five decades.

4. According to the Convention on the Law of the Sea by the United Nations (UN), apart from a 12-mile zone stretching from shorelines, which constitutes territorial water with national sovereignty, all maritime surfaces are regarded as anarchic, where ships of any states enjoy the right of 'innocent passage'.

5. For instance, Slack et al. (1996) found that, throughout the late 1980s and the 1990s, a profit of only US$0.40 per TEU per year were realized by a sample of 11 major shipping lines around the world.

6. The market here includes the Hamburg–Le Havre range and the Baltic region, covering Scandinavia, Russia, the Baltic States, Poland and the German Baltic ports.

7. According to Hayuth and Fleming (1994), centrality was generated from locally generated demands (both urban and industrial), while intermediacy was generated from port *en route* which usually involved long-distance in-transit and trans-shipment traffic.

3 Global Supply Chains and Trade Logistics: From Firm-Focal to Port-Focal

1. In 2005, the organization became the Council of Supply Chain Management Professionals. For further details, see Council of Supply Chain Management Professionals (n.d.).
2. *Jidoka* means autonomous adaptation, or 'automation with a human touch.'
3. Of course, this applies only as long as ports are viewed as firms.
4. In 2012, Singapore and Hong Kong ranked 2nd and 3rd in the world in terms of container throughputs, respectively. For further details, see Containerisation International (2013).

4 Logistics, Supply Chain and Port Evolution

1. According to anecdotal information, even if it was left idle, a ship with a capacity of 8,000 TEUs or more required at least a few million US dollars to maintain annually.
2. The strategies between hub and feeder ports could be different as their major customers could have different considerations when making port choice decisions. For example, Chang et al. (2008) argued that when making port choices, shipping lines which operated trans-continental services mainly focused on ports' marketability and service quality, while those operating feeder services emphasized market size and operational costs. Also, Ng and Kee (2008) argued that shipping lines which operated feeder services were usually less able to pressurize ports to adhere to their requests due to smaller cargo volumes. Thus, local demands were pivotal in their port choice decisions. Talley and Ng (2013) even argue that, very recently, there have been a few cases where ports have started to choose ships.
3. One should note that the management of respective ports are still the core functions of the corporatized port authorities, rather than transforming themselves into conventional private firms with investments in (foreign) ports terminals like Singapore. Indeed, PSA's traditional function was picked up by a new public authority, namely the Maritime and Port Authority of Singapore (MPA). For further details of the development of the ports of Busan and Rotteram, see Chapter 6. For the development of the ports of Piraeus and Tianjin, see Ng and Pallis (2010) and Shou et al. (2011), respectively.

4. ERS was initially a joint venture between Maersk and P&O Nedlloyd. It came under Maersk's complete control after the A.P. Moller Group (APMG) acquired P&O Nedlloyd in 2006.

5 Port and Trade Industrial Organization

1. For a thorough and rigorous characterization of regularity conditions, see Diewert (1971).
2. Illustrative examples include Aschauer (1989), Nadiri and Mamuneas (1994), Holtz-Eakin (1994) and Demetriades and Mamuneas (2000).
3. For further explanations on the Cobb-Douglas production function, see Douglas (1976).
4. The basic DEA uses deterministic linear optimization methods to construct a non-parametric piece-wise linear frontier, which can be estimated from empirical data on outputs and inputs. Although the piece-wise-linear frontier estimation was first proposed by Farrel (1957), the formal method was actually first developed by Charnes *et al.* (1978); it was input based and assumed to be CRS.

6 Government Policies and the Role of Institutions

1. For a detailed explanation on Figure 6.1 and the reform process, see Ng and Pallis (2010) and Shou et al. (2011).
2. Indeed, this was the case for many ports located between the Hamburg-Le Havre range, as well as Scandinavian ports.
3. Apart from the main reasons illustrated above, the issue of power could not be ignored. Some actors might favor certain reform options due to their vested interests or because of their political ideology, while some were more powerful than others in pushing forward certain reform options. For instance, during the process of reform in Rotterdam several options were in fact considered. In particular, the local ruling Labor Party was not in favor of all out privatization or corporatization, but preferred to keep the port authority intact as a municipal department, with some extra mandates. Indeed, it was only after the Labor Party lost the general elections that corporatization advanced to the top of the political agenda. In fact, both the city and the port authority were in favor of reform, but the aforementioned local politics impeded drastic reform.
4. The competitiveness of Dutch ports, and in particular the port of Rotterdam, had been a core part of the economic development. This was because of the remarkable size of the cluster associated with one of the biggest European ports that was key for the economy (cf. de Langen, 2002), and the need to serve another core part of the Dutch economy, i.e. the exports of manufacturing.

5. For further details on the recent international strategies of PoR, see Dooms et al. (2013).
6. The non-executive board of PoR had actually suspended a CEO from his duty in 2004 when the then CEO, Willem K. Scholten, was involved in an investigation into the underwriting of bank loans by the port. The loans, totalling 100 million Euros (about 133 million US dollars), were granted by the CEO to RDM, a group of companies mainly involved in ship repairs and maintenance, without consulting members of the executive board, the non-executive board, or the municipal board of the port of Rotterdam.
7. In fact, according to Vries (2005), the 'global' vision of the port of Rotterdam could be dated back to the nineteenth century, or even earlier.

7   Case Study – India

1. Another incentive for the participation of dry port users in dry port operation was related to the common practice in this trade, where most dry ports operators would charge their fees in advance from the users, mostly comprised of shipping lines and non-vessel owning common carriers (NVOCC). It was not uncommon for users to keep a certain amount of money with the operators and replenish the deposits at regular intervals. On the other hand, the dry port operators employed vendors for transportation and the handling of cargoes and paid them at a later date. The operators often used the surplus cash as 'seed capital' for expansion and servicing of debts. This served as a reason why many users would like to enter into the business of inland logistics even though the profit margin was thin, since cash flow would be steady and rather low risk.
2. This problem was slightly relieved by the Indian government in permitting leasing companies to import container wagons increasing the capacity of wheel and axle plants. Nevertheless, the capital-intensive nature of rail transport and the dominant position of CONCOR has ensured that this problem is unlikely to be completely addressed in the foreseeable future. Indeed, this has also reflected the dualistic nature of Indian policies in ensuring the survival of foreign investors, while at the same time restricting them to peripheral, rather than dominating, players.
3. Data from industrial sources revealed that over 98 percent of the cargo movements by road to the port of Tuticorin, in spite of the fact that the port of Cochin was closer. This was due to the Tamilnadu government's policy of promoting Tuticorin which was located within Tamilnadu, while Cochin was located in Kerala. This was achieved through offering subsidies to the road transporters so as to prevent the loss of revenues by the Tamilnadu state government. The port of Chennai though located in the same state did not get many cargoes from Tirupur due to longer distance.
4. The Indian government sometimes supported foreign operators by providing aid similar to what had been provided to state-owned corporations. However, according anecdotal information, the amount of such

aid was by no means comparable to state-owned corporations. This policy further reflected the dualistic nature of the Indian policy, where foreign investors would be tolerated (or even supported), but at the same time were restricted from increasing market power.

5. Apart from that, CONCOR also developed its own gateway terminals in its dry port near JNP in offering a single vendor service to meet the logistics requirements of customers and they are currently exploring the possibilities of entering into freight contracts with cargo interests directly to meet all the logistics needs of their customers.

6. While unable to release concrete performance data here due to confidentiality issues, according to most interviewees, until now, few foreign-based dry port operators were making any real profits since starting dry port operations in India.

8 Case Study – Brazil

1. For further details on the development and institutional reforms of Brazilian ports in the past two decades, see Galvão et al. (2013).

2. According to some interviewees, this was the case even during the peak seasons and thus during times of congestion.

3. Secondary areas imply any parts within the national territory except seaports, airports and cross-border checkpoints (the primary areas).

4. For instance, there was a debate concerning water and sanitation services where similar questions over the concession process (requirement of tender) and the appropriate jurisdiction (municipal or state level) were being discussed.

5. One should note that strikes were also common within other sectors, and affected areas such as public transportation, banking, post offices and the police.

6. This phenomenon was not restricted to Brazil. In fact, the difficulty in achieving the threshold requirement for rail service was very common among many developing economies, with India being another illustrative example (see Chapter 7).

7. The legislation concerning multimodal transport operations comprises mainly Law 9.611/98 (about Multimodal Cargo Transport); Decree 3.411/00 (regulates Law 9.611/98); Decree 5.276/04 (amended Articles 2 and 3 of Decree 3.411/00); Decree 1.563/95 (about the implementation of the Partial Agreement for the Facilitation of Multimodal Transport between Brazil, Argentina, Paraguay and Uruguay, of December 30, 1994); and Resolution 794/04 (about the accreditation of Multimodal Transport Operators).

8. In this chapter, cabotage can be understood as domestic sea cargo transportation between Brazilian ports. For further information regarding the term's definition and usage, see Taaffe et al. (1996).

9. Established in 2005 with 85,000 m^2, the dry port of Libraport was part of the Libra Group, a Brazilian company partly owned by the Mitsui group.

It controlled a wide range of logistical facilities in the state of Sao Paulo, including several port terminals. For further details, see Padilha and Ng (2012).

10. ITRI-Rodoferrovia e Serviços Ltda. is a road-rail transporting service provider and OTM in Brazil.

11. With a population of 20 million, the metropolitan area corresponded to about 50% of the state population, 3 percent of the state area and 50 percent of the GDP of the State of Sao Paulo.

12. Although the city continued to attract people and business, surrounding cities had grown at a faster rate since the 1970s as social and environmental problems led people and business out of San Paulo in the search for lower prices, fresher air and safer locations. For further details, see Padilha and Ng (2012).

13. According to Bogus and Pasternak (2004), as many as 30 percent of the shanty towns in the city of Sao Paulo were located along expressways.

9   Port-Focal Logistics: The Ideal for Future Global Supply Chains?

1. According to anecdotal information collected by Ng and Gujar (2009), such an objective also created disincentives for the state government of Tamilnadu to actively address the problem of long queues in the collection of cross-border taxes between Kerala and Tamilnadu.

2. For a detailed discussion on the potential uncertainties that ports need to tackle when planning and adapting to the impacts posed by climate change, see Ng et al. (2013a and 2013b).

References

ABEPRA, Associação Brasileira dos Portos Secos, *Sobre Portos Secos*, www.slide-share.net/datasul/sobre-portos-secos, accessed January 2010.

Abramovitz, M. (1956) 'Resource and output trends in the United States since 1870,' *American Economic Review*, 46, 5–23.

Acemoglu, D. and J. Robinson (2008) 'The role of institutions in growth and development,' Working Paper No. 10, Commission on Growth and Development, The World Bank Group, Washington, DC.

Agencia Senado (2010). Senado pode mudar lei dos portos secos, July 9, 2010, www.senado.gov.br, accessed August 15, 2010.

Aigner, A., C.A.K. Lovell and P. Schmidt (1977) 'Formulation and estimation of stochastic frontier production function models,' *Journal of Econometrics*, 86, 21–37.

Airriess, C.A. (2001) 'The regionalization of Hutchison port holdings in Mainland China,' *Journal of Transport Geography*, 9, 267–278.

Aldridge, H.E. (1999) *Organisations Evolving*, London: Sage.

Anderson, J.E. and E. van Wincoop (2003) 'Gravity with gravitas: a solution to the border puzzle,' *American Economic Review*, 93(1), 170–192.

Araujo Jr., J.T. (2004) *Condutas anticompetitivas em industrias de rede: O caso do Porto de Santos*, Ecostrat Consultores, October, available at http://www.ecostrat.net/files/Condutas_Anticompetitivas_Em_Industrias_de_Rede.pdf.

Arrow, K.J. and G. Debreu (1954) 'Existence of an equilibrium for a competitive economy,' *Econometrica*, 22, 265–290.

Arrow, A.J., H.B. Cheney, B.S. Minhas and R.W. Solow (1961) 'Capital-labor substitution and economic efficiency,' *Review of Economics and Statistics*, 63, 225–250.

Aschauer, D.A. (1989) 'Is public expenditure productive?,' *Journal of Monetary Economics*, 23, 177–200.

Assumpção, M.R.P., A. Alves and L.T. Robles (2009) *Implantação de Áreas dedicadas a Regimes Aduaneiros Especiais na Região Metropolitana da Baixada Santista: Zonas de Processamento de Exportação e Centro de Logística e Indústrias Alfandegadas*, XII SEMEAD, Universidade de Sao Paulo.

Baird, A.J. (2002) 'The economics of container trans-shipment in Northern Europe,' *International Journal of Maritime Economics*, 4(3), 249–280.

Baird, A.J. (2004) 'Trans-shipment: hub port selection,' *Cargo Systems*, May 2004, 44–47.

Bajaj, V. (2010) 'India's clogged rail lines stall economic progress' *The New York Times*, June 16, B1, B4.

Baum-Snow, N. (2007) 'Did highways cause suburbanization?,' *Quarterly Journal of Economics*, 122(2), 775–805.

Beattie, B.R. and C.R. Taylor (1985) *The Economics of Production*, New York, NY: John Wiley & Sons (Reprint edition in 1993 by Krieger Publishing Co., Malabar, FL).

Behrans, K., A.R. Lamorgese, G.I.P. Ottaviaon and T. Tabuchi (2007) 'Changes in transport and non-transport costs: local vs. global impacts in a spatial network,' *Regional Science and Urban Economics*, 37, 625–648.

Bekemans, L. and S. Beckwith (1996) *Ports for Europe: Europe's Maritime Future in a Changing Environment*, Brussels: European Interuniversity Press.

Bennathan, E. and A.A. Walters (1979) *Port Pricing and Investment Policy for Developing Countries*, Oxford: Oxford University Press.

Bergantino, A.S. and A.W. Veenstra (2002) 'Interconnection and co-ordination: an application of network theory to liner shipping,' *International Journal of Maritime Economics*, 4, 231–248.

Bichou, K. and R. Gray (2004) 'A logistics and supply chain management approach to port performance measurement,' *Maritime Policy & Management*, 31(1), 47–67.

Bird, J.H. (1971) *Seaports and Seaport Terminals*, London: Hutchison.

Bogus, L. and S. Pasternak (2004) A cidade dos extremos. XIV Encontro Nacional de Estudos Populacionais. ABEP, September 20–24, 2004.

BPA (2006) White Paper on Busan Port Authority: The Establishment, Busan: BPA.

BPA website: www.busanpa.com, date accessed May 2013.

Braga, V. (2008) Logística, planejamento territorial dos transportes e o projeto dos Centros Logísticos Integrados no Estado de São Paulo. e-Premissas, Revista de Estudos Estratégicos, No. 3, Jan/Jun.

Brenner, N. (1998) 'Between fixity and motion: accumulation, territorial organization and the historical geography of spatial scales,' *Environment & Planning D*, 16(4), 459–481.

Brooks, M.R. and K. Cullinane (2007) *Devolution, Port Governance and Port Performance*, London: Elsevier.

Buchanan, J.M. (1964) 'Is economics the science of choice?,' *Roads to Freedom: Essays in Honour of Friedrick A. von Hayek*, London: Routledge/Kegan Paul, pp. 47–64.

Buchanan, J.M. (1975) 'A contractarian paradigm for applying economics theory,' *American Economic Review*, 65(May), 225–230.

Buitelaar, E., A. Lagendijk and W. Jacobs (2007) 'A theory of institutional change: illustrated by Dutch city-provinces and Dutch land policy,' *Environment and Planning A*, 39(4), 891–908.

Cachon, G.P. (2003) 'Supply chain coordination with contracts,' in A.G. de Kok and S.C. Graves (eds) *Handbooks in Operations Research and Management Science, 11: Supply Chain Management: Design, Coordination and Operation*, Boston: Elsevier, pp. 229–340.

Cachon, G.P. and L.A. Lariviere (2005) 'Supply chain coordination with revenue-sharing contracts: strengths and limitations,' *Management Science*, 51(1), 30–44.

Carbone, V. and M. Martino (2003) 'The changing role of ports in supply-chain management: an empirical analysis,' *Maritime Policy & Management*, 30(4), 305–320.

Campbell, S. (1993) 'Increasing trade, declining port cities: port containerization and the regional diffusion of economic benefits,' in H. Noponen, J. Graham and A.R. Markusen (eds) *Trading Industries, Trading Regions*, New York: Guilford, pp. 212–227.

Camps, T.W.A. (1996) *Is er nog toekmst nah et sectorenmodel: Flexibiliteit van gemeentelijke organisaties*, Velp: Rijnconsult Group.

Cass, S. (1998) *World Port Privatisation: Finance, Funding and Ownership: A Cargo Systems Report*, London: IIR.

Chang, Y.T., S.Y. Lee and J.L. Tongzon (2008) 'Port selection factors by shipping lines: different perspectives between trunk liners and feeder service providers,' *Marine Policy*, 32(6), 877–885.

Chapman, K. and D.F. Walker (1991) *Industrial Location* (2nd edn), Oxford: Blackwell.

Charnes, A., W.W. Cooper and E. Rhodes (1978) 'Measuring the efficiency of decision making units,' *European Journal of Operational Research*, 2, 429–444.

Chen, A. and N. Groenewold (2011) 'Regional equality and national development in China: is there a trade-off?,' *Growth and Change*, 42(4), 628–669.

Chilcote, P.W. (1988) 'The containerization story: meeting the competition in trade' in M.J. Hershman (ed.) *Urban Ports and Harbour Management*, London: Taylor and Francis, pp. 125–146.

Chlomoudis, C.I. and A.A. Pallis (2002), *European Port Policy: Towards a Long-term Strategy*, Cheltenham: Edward Elgar.

Chrzanowski, I.H. (1975) *Concentration and Centralisation of Capital in Shipping*, Westmead: Saxon House, D.C. Heath.

Clarke, X., D. Dollar and A. Micco (2004) 'Port efficiency, maritime transport costs, and bilateral trade,' *Journal of Development Economics*, 75, 417–450.

Coase, R.H. (1992) 'The institutional structure of production,' *American Economic Review*, 82(4), 713–719.

Coelli, T., D.P. Rao and G.E. Battese (1998) *An Introduction to Efficiency and Productivity Analysis*, Cambridge: Cambridge University Press.

Cogan, J.F. (1981) 'Fixed costs and labor supply,' *Econometrica*, 49, 945–963.

CONCOR website: www.concorindia.com, accessed on February 2009.

Containerisation International website: http://www.lloydslist.com/ll/sector/containers, accessed August 2013.

Containerization International Yearbooks, 1997–2004.

Costinot, A., D. Donaldson and I. Komunjer (2012) 'What goods do countries trade? a quantitative exploration of Ricardo's ideas,' *The Review of Economic Studies*, 79(2), 581–608.

Council of Supply Chain Management Professionals website: http://www.cscmp.org, accessed September 2013.

Cullinane, K. and M. Khanna (2000) 'Economies of scale in large container ships,' *Journal of Transport Economics and Policy*, 33(2), 185–208.

Cullinane, K. and D.W. Song (2002) 'Port privatization policy and practice,' *Transport Reviews*, 22, 55–75.

Dada, M. and K. Srikanth (1987) 'Pricing policies for quantity discounts,' *Management Science*, 33(10), 1247–1252.

Damas, P. (2001) 'Tranship or direct – a real choice,' *American Shippers*, June 2001, 56–60.

d'Aunno, T., M. Succi and J.A. Alexander (2000) 'The role of institutional and market forces in divergent organisational change,' *Administrative Science Quarterly*, 45(4), 679–698.

de Langen, P.W. (1998) 'The future of small and medium sized ports,' in G. Sciutto and C.A. Brebbia (eds) *Maritime Engineering and Ports*, Southampton: WIT Press, pp. 263–279.

de Langen, P.W. (2002) 'Clustering and performance: the case of maritime clustering in the Netherlands,' *Maritime Policy and Management*, 29(3), 209–221.

de Langen, P.W. and A.A Pallis (2007) 'Entry barriers in seaports,' *Maritime Policy and Management*, 34(5), 427–440.

de Lombaerde, P. and A. Verbeke (1989) 'Assessing international seaport competition: a tool for strategic decision-making,' *International Journal of Transport Economics*, XVI(2), 176–192.

Demetriades, P. and T.P. Mamuneas (2000) 'Intertemporal output and employment effects of public infrastructure capital: evidence from 12 OECD economies,' *Economic Journal*, 110(465), 687–712.

Denzau, A.T. and D. North (1994) 'Shared mental models: ideologies and institutions,' *Kyklos*, 47(1), 3–31.

Diewert, W.E. (1971) 'An application of the Shephard duality theorem: a generalized Leontief production function,' *Journal of Political Economy*, 79(3), 481–507.

Dixit, A.K. and J. Stiglitz (1977) 'Monopolistic competition and optimum product diversity,' *The American Economic Review*, 67(3), 297–308.

Do, N.H., K.C. Nam and Q.M. Ngoc Le (2011) 'A consideration for developing a dry port system in Indochina area,' *Maritime Policy and Management*, 38(1), 1–9.

Donaldson, D. (2010) 'Railroad of the Raj: estimating the impact of transportation infrastructure,' working paper, The National Bureau of Economic Research (NBER), Cambridge, MA (no. 16487).

Dooms, M., L. van der Lugt and P.W. de Langen (2013) 'International strategies of port authorities: the case of the Port of Rotterdam Authority,' *Research in Transportation Business and Management*, 8, 148–157.

Douglas, P.H. (1976): 'The Cobb-Douglas production function once again: its history, its testing, and some new empirical values,' *Journal of Political Economy*, 84(5), 903–916.

Ducruet, C. (2007) 'A meta-geography of port-city relationships,' in J. Wang, D. Olivier, T. Notteboom, and B. Slack (eds): *Ports, Cities and Global Supply Chains*, Ashgate, Aldershot, pp. 157–172.

Ducruet, C. and I. Lugo (2013) 'Cites and transport networks in shipping and logistics research', *The Asian Journal of Shipping and Logistics*, 29(2), 145–166.

Dunning, J.H. (2000) 'Globalisation and the new geography of foreign direct investment,' in N. Woods (ed.) *The Political Economy of Globalisation*, London: Macmillan, pp. 20–53.

Duranton, G., P. Morrow and M. Turner (2011) 'The fundamental law of road congestion: evidence from the US,' *American Economic Review*, 101(6), 2616–2652.

Eaton, J. and S. Kortum (2002) 'Technology, geography, and trade,' *Econometrica*, 70(5), 1741–1779.

Ekberg, E., E. Lange and E. Merok (2012) 'Building the networks of trade: perspectives on twentieth-century maritime history,' in G. Harlaftis, S. Tenold and J.M. Valdaliso (eds) *World's Key Industry: History and Economics of International Shipping*, Basingstoke: Palgrave Macmillan, pp. 88–105.

Emmons, H. and S.M. Gilbert (1998) 'Note: the role of returns policies in pricing and inventory decision of catalogue goods,' *Management Science*, 44(3), 276–283.

Farrel, M.J. (1957) 'The measurement of productive frontier,' *Journal of the Royal Statistical Society: Series A*, CXX(3), 253–290.

Fernandez-Alles, M.D.L.L. and R. Llamas-Sanchez (2008) 'The neo-institutional analysis of change in public services,' *Journal of Change Management*, 8(1), 3–20.

Fleming, D.K. and Y. Hayuth (1994) 'Spatial characteristics of transportation hubs: centrality and intermediacy,' *Journal of Transport Geography*, 2(1), 3–18.

Fogel, R.W. (1964) *Railroads and American Economic Growth: Essays in Economic History*, Baltimore: Johns Hopkins University Press.

Folha (2009). *Outro lado: Receita diz que sistema será reformulado*, October 13, 2009, www.folha.uol.com.br, date accessed March 16, 2010.

Forrester, J.W. (1961) *Industrial Dynamics*, Cambridge, MA: MIT Press.

Fujita, M. and T. Mori (1996) 'The role of ports in the making of major cities: self-agglomeration and hub effects', *Journal of Development Economics*, 49(1), 93–120.

Galvão, C.B., L.T. Robles and L.C. Guerise (2013) 'The Brazilian seaport system: a post-1990 institutional and economic review,' *Research in Transportation Business and Management*, 8, 17–29.

Gérard Cachon, Martin Lariviere (2005) 'Supply chain coordination with revenue sharing: strengths and limitations,' *Management Science*, 51(1), 30–44.

Gertler, M.S. (2001) 'Best practice? Geography, learning and the institutional limits to strong convergence,' *Journal of Economic Geography*, 1, 5–26.

Goldberg, D.J.K. (2009) Regulação do setor portuário no Brasil: Análise do novo modelo de concessão de portos organizados, MSc Dissertation, Universidade de Sao Paulo.

Government of India (2003) *The Gazette of India: Extraordinary (Part II – Section 1)*, No. 12, New Delhi: Government of the Republic of India.

Granovetter, M. (1985) 'Economic action and social structure: the problem of embeddedness,' *American Journal of Sociology*, 91(3), 481–510.

Grant, W. (1997) 'Perspectives on globalization and economic coordination,' in J.R. Hollingsworth and R. Boyer (eds) *Contemporary Capitalism: The Embeddedness of Institutions*, Cambridge: Cambridge University Press, pp. 319–336.

Gubbins, E.J. (1988) *Managing Transport Operations*, London: Kogan Page.

Ha, M.S. (2003) 'A comparison of service quality at major container ports: implications for Korean ports,' *Journal of Transport Geography*, 11, 131–137.

Hall, P.A. (1986) *Governing the Economy: The Politics of State Intervention in Britain and France*, Cambridge: Polity.

Hall, P.A. and R.C.R. Taylor (1998) 'Political science and the three new institutionalisms,' in K. Soltan, E. Soltan and E.M. Uslaner (eds) *Institutions and Social Order*, Virginia Haufler: University of Michigan Press, pp. 14–44.

Hall, P.V. (2003) 'Regional institutional convergence? Reflections from the Baltimore Waterfront,' *Economic Geography*, 79(3), 347–363.

Hall, R.E. and C.I. Jones (1999) 'Why do some countries produce so much more output per worker than others?,' *Quarterly Journal of Economics*, 114(1), 83–116.

Hanson, S. (2000) 'Transportation: hooked on speed, eyeing sustainability,' in E. Sheppard and T.J. Barnes (eds) *A Companion to Economic Geography*, Oxford: Blackwell.

Haralambides, H.E. (2000) 'A second scenario on the future of the hub-and-spoke system in liner shipping,' Paper presented in the *Latin Ports and Shipping 2000 Conference*, Lloyd's List, November 14–16, Miami, FL, US.

Haralambides, H.E. (2005) 'Bigger isn't necessarily better,' *Lloyd's List: Letter to the Editor*, London: Lloyd's List.

Haralambides, H.E. and R. Behrens (2000) 'Port restructuring in a global economy: an Indian perspective,' *International Journal of Transport Economics*, 27(1), 19–39.

Hariharan, K.V. (2004) *Containerization and Multimodal Transport in India*, Mumbai: Shroff.

Harlaftis, G., S. Tenold and J.M. Valdaliso (eds) (2012) *World's Key Industry: History and Economics of International Shipping*, New York: Palgrave Macmillan.

Harley, C.K. (2012) 'Building the networks of trade: perspectives on twentieth-century maritime history,' in G. Harlaftis, S. Tenold and J.M. Valdaliso (eds) *World's Key Industry: History and Economics of International Shipping*, New York: Palgrave Macmillan, pp. 29–42.

Harvey, D. (2005) *A Brief History of Neoliberalism*, Oxford: Oxford University Press.

Hayuth, Y. (1981) 'Containerization and the load center concept,' *Economic Geography*, 57(2), 160–176.

Hayuth, Y. and D.K. Fleming (1994) 'Concepts of strategic commercial location: the case of container ports,' *Maritime Policy and Management*, 21(3), 187–193.

Hayuth, Y. and D. Hilling (1992) 'Technological change and seaport development,' in B. Hoyle and D. Pinder (eds) *European Port Cities in Transition*, London: Belhaven, pp. 4–58.

Heaver, T.D. (1993) 'Shipping and the market for port services,' in G. Blauwens, G. de Brabander and E. van de Voorde (eds) *De Dynamiek Van Een Haven*, Kapellen: Uitgeverij Pelckmans, pp. 227–248.

Heaver, T.D. (1995) 'The implications of increased competition among ports for port policy and management,' *Maritime Policy & Management*, 22(2), 125–133.

Heaver, T.D. (2002) 'The evolving roles of shipping lines in international logistics,' *International Journal of Maritime Economics*, 4(3), 210–230.

Henderson, J., P. Dicken, M. Hess, N. Coe and H.W.C. Yeung (2002) 'Global production networks and the analysis of economic development,' *Review of International Political Economy*, 9(3), 436–464.

Hendriks, F. and P. Tops (1999) 'Between democracy and efficiency: trends in local government in the Netherlands and Germany,' *Public Administration*, 77(1), 133–153.

Hesse, M. and J.P. Rodrigue (2004), 'The transport geography of logistics and freight distribution,' *Journal of Transport Geography*, 12(3), 171–184.

Hinz, C. (1996) 'Prospects for a European ports policy: a German view,' *Maritime Policy & Management*, 23(4), 337–340.

Hijjar, M.F. and F.M.B. Alexim (2006) Avaliação do acesso aos terminais portuários e ferroviários de contêineres no Brasil. Coppead/UFRJ, Centro de Estudos em Logística, accessible at: www.centrodelogistica.com.br/new/fs-panorama_logistico3.htm.

Holtz-Eakin, D. (1994) 'Public sector capital and the productivity puzzle,' *Review of Economics and Statistics*, 76, 12–21.

Homosombat, W., A.K.Y. Ng and X. Fu (forthcoming) 'Regional transformation and port cluster competition: the case of the Pearl River Delta in South China,' *Growth and Change*.

Hoover, E.M. and F. Giarratani (1985) *An Introduction to Regional Economics* (3rd edn), New York: McGraw-Hill.

Hotelling, H. (1929) 'Stability of competition,' *Economic Journal*, 39, 41–57.

Hoyle, B.S. (1989) 'The port-city interface: Trends problems and examples,' *Geoforum*, 20(4), 429–435.

Huybrechts, M., H. Meersman, E. van de Voorde, E. van Hooydonk, A. Verbeke and W. Winkelmans (2002) *Port Competitiveness: An Economic and Legal Analysis of the Factors Determining the Competitiveness of Seaports*, Antwerp: De Boeck.

Hyundai (2005) *Presentation of Mega Container Carrier: Korean Yard Now Accepting Orders*, Ulsan: Hyundai Heavy Industries.

IDB (2013) 'Port-centric development: strategic logistics investments', Technical note, Department of Infrastructure and Investment, Inter-American Development Bank (no. No. IDB-TN-510).

Investment Commission of the Government of India (2006) *Investment Strategy for India*. Investment Commission report submitted to the Government of India, E/ESCAP/MCT/SGO/8 February.

Islam, D.M.Z., J. Dinwoodie and M. Roe (2005) 'Towards supply chain integration through multimodal transport in developing economies: the case of Bangladesh,' *Maritime Economics and Logistics*, 7, 382–399.

Jacobs, W. and P.V. Hall (2007) 'What conditions supply chain strategies of ports? The case of Dubai,' *Geojournal*, 68, 327–342.

Jansson, J.O. and D. Shneerson (1982) 'The optimal ship size,' *Journal of Transport Economics and Policy*, 16(3), 217–238.

Jessop, B. and S. Oosterlynk (2008) 'Cultural political economy: on making the cultural turn without falling into soft economic sociology,' *Geoforum*, 39(3), 1155–1169.

Johnson, K.M. and H.C. Garnett (1971) *The Economics of Containerisation*, London: George Allen & Unwin.

Johnson, G., S. Smith and B. Codling (2000) 'Microprocesses of institutional change on the context of privatisation,' *Academy of Management Review*, 25(3), 572–581.

Juhel, M.H. (2001) 'Globalisation, privatisation and restructuring of ports,' *International Journal of Maritime Economics*, 3, 139–174.

Junior, G.A.D.S., A.K. Beresford and S.J. Petit (2003) 'Liner shipping companies and terminal operators: internationalisation or globalisation?,' *Maritime Economics & Logistics*, 5, 393–412.

Kaukiainen, Y. (2012) 'The advantages of water carriage: scale economies and shipping technology, c. 1870–2000,' in G. Harlaftis, S. Tenold and J.M. Valdaliso (eds) *World's Key Industry: History and Economics of International Shipping*, New York: Palgrave Macmillan, pp. 64–87.

Kim, S.H., M.A. Cohen and S. Netessine (2010) 'Reliability or inventory? Analysis of product support contracts in the defense industry,' Working paper, Yale University, New Haven, CT.

Keedi, S. (2010). *Containers na Ferrovia*. Revista Sem Fronteiras, February 2010.

Kendrick, J. (1961) *Productivity Trends in the United States*, Princeton, NJ: Princeton University Press.

Kingdon, J.W. (1995) *Agendas, Alternatives and Public Policies* (2nd edn), New York: Longman.

Knox, B. (2001). Reassessing the impact of institutions on economic reform in Brazil, master thesis, University of Florida.

Krugman, P. (1998) 'What's new about the new economic geography?,' *Oxford Review of Economic Policy*, 14(2), 7–17.

Kumar, S. (2000) 'An evaluation of liner strategies in the context of contemporary supply chain management practices,' *Journal of Transport Management*, 12(2), 55–64.

Lacerda, S.M. (2005) 'Logística ferroviária do Porto de Santos: a integração operacional da infra-estrutura compartilhada,' Revista BNDESS, 12(24), 189–210.

Lam, J.S.L., A.K.Y. Ng and X. Fu (2013) 'Stakeholder management for establishing sustainable regional port governance,' *Research in Transportation Business and Management*, 8, 30–38.

Lariviere, M.A. (1999) 'Supply chain contracting and coordination with stochastic demand,' in S. Tayur, R. Ganeshan and M. Magazine (eds)

Quantitative Models for Supply Chain Management, Norwell, MA: Kluwers, pp. 234–268.

Lau, Y.Y., A.K.Y. Ng, X. Fu and K.X. Li (2013) 'Evolution and research trends of shipping,' *Maritime Policy & Management*, 40(7), 654–674.

Lee, C.H. (2001) 'Coordinated stocking, clearance sale, and return policies for a supply chain,' *European Journal of Operational Research*, 131(3), 491–513.

Lee, H.L., V. Padmanabhan and S. Whang (1997) 'The bullwhip effect in supply chains', *Sloan Management Review*, 38(3), 93–102.

Lee, S.W., D.W. Song and C. Ducruet (2008) 'A tale of Asia's world ports: the spatial evolution in global hub port cities,' *Geoforum*, 39(1), 372–385.

Limão, N. and A. Venables (2001) 'Infrastructure, geographical disadvantage, transport costs and trade,' *World Bank Economic Review*, 15(3), 451–479.

Lin, G.C.S. (2010) 'Understanding land development problems in globalizing China,' *Eurasian Geography and Economics*, 51(1), 80–103.

Liu, Z. (1995) 'Ownership and productive efficiency: the experience of British ports,' in J. McConville and J. Sheldrake (eds) *Transport in Transition: Aspects of British and European Experience*, Aldershot: Avebury, pp. 163–182.

Liu, J.J. (2009) 'Respond to the test of uncertain global disruptions: port-integrated logistics and supply chain', Keynote speech delivered during the International Forum on Shipping, Ports and Airports (IFSPA) 2009, Hong Kong, China, 24–27 May, accessible at: http://www.icms.polyu.edu.hk/ifspa2009/IFSPA2009%20Conference%20pdf/Industrial-Forum/IndustrialForum_1-Welcome-Speech-JohnLiu.pdf.

Liu, J.J. (ed.) (2011) *Supply Chain and Transport Logistics*, London: Routledge.

Maasvlakte 2 website: http://www.maasvlakte2.com, accessed August 2013.

Majumdar, B. (2012) 'Port-centric logistics: providing competitive advantage for gateway ports', *Port Technology International*, 53, 25–26.

Mangan, J., C. Lalwani and B. Fynes (2008) 'Port-centric logistics', *The International Journal of Logistics Management*, 19(1), 29–41.

March, J.G. and J.P. Olsen (1989) *Rediscovering Institutions: The Organisational Basis of Politics*, New York: The Free Press.

Martin, J. and B.J. Thomas (2001) 'The container terminal community,' *Maritime Policy & Management*, 28(3), 279–292.

Martinsons, M.G. (2002) 'Electronic commerce in China: emerging success stories,' *Information & Management*, 39(7), 571–579.

Matayoshi, N.N. (2004) *O instituto da permissão de serviços públicos no atual ordenamento jurídico brasileiro*, Monograph, Tribunal de Condas da União, Instituto Serzedello Corrêa, Brasília.

McCalla, R.J. (1999) 'Global change, local pain: intermodal seaport terminals and their service areas,' *Journal of Transport Geography*, 7(4), 247–254.

McFadden, D. (1963) 'Constant elasticity of substitution production functions,' *Review of Economic Studies*, 30, 73–83.

Meersman, H. and van de Voorde, E. (1998) 'Coping with port competition in Europe: a state of the art,' in G. Sciutto and C.A. Brebbia (eds) *Maritime Engineering and Ports*, Southampton: WIT, pp. 281–290.

Meersman, H., E. van de Voorde, and T. Vanelslander (2005) 'Ports as hubs in the logistics chain', in H. Leggate, J. McConville and A. Morvillo (eds) *International Maritime Transport: Perspectives*, London: Routledge, pp. 123–129.

Melitz, M. and D. Trefler (2012) 'Gains from trade when firms matter,' *Journal of Economic Perspectives*, 26(2), 91–118.

Michaels, G. (2008) 'The effect of trade on the demand for skill – evidence from the interstate highway system,' *Review of Economics and Statistics*, 90(4), 683–701.

Ministry of Shipping of the Government of India website: www.shipping.gov.in, date accessed July 2013.

Miyashita, K. (2005) 'The logistics strategy of Japanese ports: the case of Kobe and Osaka,' in P.T.W. Lee and K. Cullinane (eds) *World Shipping and Port Development*, New York: Palgrave Macmillan, pp. 181–198.

MOMAF (2004) *Materials in Explaining the Establishment of BPA. Official briefings to the Media*, Seoul: National Government of the Korean Republic.

Monios, J. and G. Wilmsmeier (2012) 'Port-centric logistics, dry ports and offshore logistics hubs: strategies to overcome double peripherality?', *Maritime Policy & Management*, 39(2), 207–226.

Moon, C.I. (1994) 'Changing patterns of business-government relations in South Korea,' in A. MacIntyre (ed.) *Business and Government in Industrialising Asia*, St. Leonards: Allen and Unwin, pp. 142–166.

Nadiri, M.I. and T. Mamuneas (1994) 'The effects of public infrastructure and R&D capital on the cost structure and performance of the US manufacturing industries,' *Review of Economics and Statistics*, 76(1), 22–37.

Nam, H.S. and D.W. Song (2011) 'Defining maritime logistics hub and its implications for container port,' *Maritime Policy & Management*, 38(3), 269–292.

Nascimento, J.P. (2005) *Vantagens e limitações decorrentes da implantação da Lei de Modernização dos Portos*, MSc Thesis, Universidade Federal do Rio de Janeiro.

Ng, A.K.Y. (2006) *Theory and Structure of Port Competition: A Case Study of Container Transhipment in North Europe.* DPhil thesis, University of Oxford.

Ng, A.K.Y. (2009) *Port Competition: The Case of North Europe*, Saarbrucken; VDM.

Ng, A.K.Y. and I.B. Cetin (2012) 'Locational characteristics of dry ports in developing economies: some lessons from Northern India', *Regional Studies*, 46(6), 757–773.

Ng, A.K.Y. and G.C. Gujar (2009) 'The spatial characteristics of inland transport hubs: evidences from Southern India,' *Journal of Transport Geography*, 17(5), 346–356.

Ng, A.K.Y. and G.C. Gujar (2009a) 'The spatial characteristics of dry ports in India', *Transport and Communications Bulletin for Asia and the Pacific*, 78, 102–111.

Ng, A.K.Y. and G.C. Gujar (2009b) 'Government policies, efficiency and competitiveness: the case of dry ports in India', *Transport Policy*, 16(5), 232–239.

Ng, A.K.Y. and J.K.Y. Kee (2008) 'The optimal ship sizes of container liner feeder services in Southeast Asia: a ship operator's perspective,' *Maritime Policy & Management*, 35(4), 353–376.

Ng, A.K.Y. and J.J. Liu (2010) 'The port and maritime industries in the post-2008 world: challenges and opportunities'. *Research in Transportation Economics*, 27(1), 1–3.

Ng, A.K.Y. and A.A. Pallis (2007a) 'Reforming port governance: the role of political culture', *Proceedings of Annual Conference of the International Association of Maritime Economists (IAME) 2007*, Athens, Greece, 4–6 July.

Ng, A.K.Y. and A.A. Pallis (2007b) 'Differentiation of port strategies in addressing proximity: the impact of political culture', *Proceedings for the International Congress on Ports in Proximity: Competition, Cooperation and Integration*, Antwerp, Belgium and Rotterdam, Netherlands, 5–7 December.

Ng, A.K.Y. and A.A. Pallis (2010) 'Port governance reforms in diversified institutional frameworks: generic solutions, implementation asymmetries', *Environment and Planning A*, 42(9), 2147–2167.

Ng, A.K.Y. and J.L. Tongzon (2010) 'The transportation sector of India's economy: dry ports as catalysts for regional development', *Eurasian Geography and Economics*, 51(5), 669–682.

Ng, A.K.Y., A. Becker and M. Fischer (2013a): 'A theoretical discussion on climate change, port adaptation strategies and institutions,' *Proceedings of the Canadian Transportation Research Forum (CTRF) 2013*, Halifax, NS, Canada, June 10–11.

Ng, A.K.Y., S.L. Chen, S. Cahoon, B. Brooks and Z. Yang (2013b): 'Climate change and the adaptation strategies of ports: the Australian experiences,' *Research in Transportation Business and Management*, 8, 186–194.

Ng, A.K.Y., F. Padilha and A.A. Pallis (2013) 'Institutions, bureaucratic and logistical roles of dry ports: the Brazilian experiences', *Journal of Transport Geography*, 27, 46–55.

North, D. (1990) *Institutions, Institutional Change and Economic Performance*, Cambridge: Cambridge University Press.

Notteboom, T.E. (2002) 'Consolidation and contestability in the European container handling industry,' *Maritime Policy & Management*, 29(3), 257–269.

Notteboom, T.E. and J.P. Rodrigue (2005) 'Port regionalization: towards a new phase in port development', *Maritime Policy & Management*, 32(3), 297–313.

Notteboom, T. and J.P. Rodrigue (2007) 'Re-assessing port-hinterland relationships in the context of global commodity chains,' in J. Wang, D. Olivier, T. Notteboom and B. Slack (eds) *Ports, Cities and Global Supply Chains*, Ashgate: Aldershot, pp. 51–66.

Notteboom, T.E. and W. Winkelmans (2001) 'Structural changes in logistics: how will port authorities face the challenge?,' *Maritime Policy & Management*, 28(1), 71–89.

Ocean Shipping Consultants (OSC) (2004) *Felixstowe South Reconfiguration: The Need for Deepwater Container Capacity*. Unpublished consulting report, Surrey: Ocean Shipping Consultants.

O Estado de São Paulo (2010) Malha ferroviária é mal aproveitada no transporte de carga. O Estado de São Paulo, May 9.

Ohno, T. (1978) *Toyota Production Systems*, Tokyo: Diamond-Sha (in Japanese).

Oliver, C. (1992) 'The antecedents of deinstitutionalization,' *Organisation Studies*, 13(4), 563–588.

Ottaviano, G.L.P. (2011) 'New' new economic geography: firm heterogeneity and agglomeration economies,' *Journal of Economic Geography*, 11(2), 231–240.

Pace, F. (2010). Director of the Syndicate of Customs Brokers of Minas Gerais. Interview on March 17, 2010.

Padilha, F. and A.K.Y. Ng (2012) 'The spatial evolution of dry ports in developing economies: the Brazilian experiences', *Maritime Economics & Logistics*, 14(1), 99–121.

Pallis, A.A. (2007) 'Whither port strategy? theory and practice in Conflict,' in A.A. Pallis (ed.) *Maritime Transport: The Greek Paradigm*, Transport Economics Series No 21, London: Elsevier, pp. 345–386

Palmer, S. (1999) 'Current port trends in an historical perspective,' *Journal for Maritime Research*, December, 1–13.

Panayides, P.M. and K. Cullinane (2002) 'Competitive advantage in liner shipping: a review and research agenda,' *International Journal of Maritime Economics*, 4, 189–209.

Parry, I.W.H. and A. Bento (2001) 'Revenue recycling and the welfare effects of road pricing,' *Scandinavian Journal of Economics*, 103(4), 645–671.

Pasternack, B. (1985) 'Optimal pricing and returns policies for perishable commodities,' *Marketing Science*, 4(2), 166–176.

Pauka, T. and R. Zunderdorp (1990) *De banana wordt bespreekbaar: Cultuurverandering in ambtelijk en politiek Groninken*, Amsterdam: Nijgh and Van Ditmar.

Peters, H.J.F. (2001) 'Developments in global seatrade and container shipping markets: their effects on the port industry and private sector involvement,' *International Journal of Maritime Economics*, 3, 3–26.

Pierson, P. (1993) 'When effects become cause: Policy feedback and political change,' *World Politics*, 45(4), 595–628.

Pimentel, C.P. (2004) História do Porto de Santos. Novo Milênio. Available at *www.novomilenio.inf.br/porto/portoh00.htm*.

PoR (2004) 'Port Vision 2020,' PoR: Rotterdam.

PoR (2005) 'PoR Business Plan, 2006–2010,' Unpublished report, Rotterdam: PoR.

PoR (2007) 'The Port of Rotterdam and her marketing and sales approach,' Unpublished article, Rotterdam: PoR.

PoR, www.portofrotterdam.com., accessed December 2013.

PoR, 'Annual Reports, 2002–2005,' Rotterdam: PoR.

Porter, M.E. (2000) 'Locations, clusters and company strategy,' in G. Clark, M.P. Feldman and M.S. Gertler (eds) *The Oxford Handbook of Economic Geography*, Oxford: Oxford University Press.

Pouder, R.W. (1996) 'Privatizing services in local government: an empirical assessment of efficiency and institutional explanations,' *Public Administration Quarterly*, 20(1), 103–127.

Powell, T. (2001) *The Principles of Transport Economics*, London: PTRC.

Powell, W. and P. di Maggio (1991) *The New Institutionalism in Organisational Analysis*, Chicago: Chicago University Press.

Prime, P.B. (2009) 'China and India enter global markets: a review of comparative economic development and future prospects,' *Eurasian Geography and Economics*, 50(6), 621–642.

Psaraftis, H.N. (1998) 'Strategies for Mediterranean port development,' in G. Sciutto and C.A. Brebbia (eds) *Maritime Engineering and Ports*, Southampton: WIT, pp. 255–262.

Pusan Newport Co. website: http://www.pncport.com, accessed December 2007.

Qiao, B., J. Martinez-Vazquez and Y. Xu (2008) 'The trade-off between growth and equity in decentralization policy: China's experience,' *Journal of Development Economics*, 86, 112–128.

Raghuram, G. (2005) 'Policy issues in port development – a case study of India's premier port,' accessible at: http://www.ficci.com.

Rahimi, M., A. Asef-Vaziri and R. Harrison (2008) 'An inland port location-allocation model for a regional intermodal goods movement system,' *Maritime Economics and Logistics*, 10(4), 362–379.

Rimmer, P.J. (1998) 'Ocean liner shipping services: corporate restructuring and port selection/competition,' *Asia Pacific Viewpoint*, 39(2), 193–208.

Rimmer, P.J. (2007) 'Port dynamics since 1965: Past patterns, current conditions and future directions,' *Journal of International Logistics and Trade*, 5(1), 75–97.

Robinson, R. (2002) 'Ports as elements in value-driven chain systems: the new paradigm,' *Maritime Policy & Management*, 29(3), 241–255.

Rodrigues, R.C.A. (2009) *Modernização dos Portos – Análise das transformações na estrutura portuária do país e dos impactos na região de Sepetiba*, 12 Encuentro de Geógrafos de América Latina, Montevideo.

Rodrigue, J.P., J. Debrie, A. Fremont, and E. Gouvernal (2010) 'Functions and actors of inland ports: European and North American dynamics,' *Journal of Transport Geography*, 18(4), 519–529.

Romer, P. (1990) 'Technical change and the aggregate production function,' *Review of Economics and Statistics*, 39, 312–320.

Roso, V. (2008) 'Factors influencing implementation of a dry port,' *International Journal of Physical Distribution and Logistics Management*, 38(10), 782–798.

Roso, V., J. Woxenius and K. Lumsden (2009) 'The dry port concept: connecting container seaports with the hinterland,' *Journal of Transport Geography*, 17(5), 338–345.

Rutten, B.C.M. (1998) 'The design of a terminal network for intermodal transport,' *Transport Logistics*, 1, 279–298.

Ryoo, D.K. and Y.S. Hur (2007) 'Busan: the future logistics hub of Northeast Asia,' in K. Cullinane and D.W. Song (eds) *Asian Container Ports*, New York: Palgrave Macmillan, pp. 34–61.

Sachs, J.D., N. Bajpai, M.F. Blaxill and A. Maira (2000) *Foreign Direct Investment in India – How can $10 Billion of Annual Inflows be Realized?* Report presented to the Ministry of Commerce and Industry of the Government of India, jointly prepared by the Centre for International Development, Harvard University and the Boston Consulting Group.

Sahay, B.S. and R. Mohan (2003) '3PL: an Indian perspective,' *International Journal of Physical Distribution and Logistics Management*, 37(7), 582–606.

Schoenberger, E. (1997) *The Cultural Crisis of the Firm*, Cambridge, MA: Blackwell.

Schumpeter, J. (1942) *Capitalism, Socialism, and Democracy*, New York, NY: Harper & Bros.

Shirley, C. and C. Winston (2004) 'Firm inventory behavior and the returns from highway infrastructure investments,' *Journal of Urban Economics*, 55, 398–415.

Shou, C., Ng, A.K.Y. and Pallis, A.A. (2011) 'Transport node governance in a changing world: the institutional reform of Tianjin port in China,' in T.E. Notteboom (ed.) *Current Issues in Shipping, Ports and Logistics*, Antwerp: University Press Antwerp, pp. 467–481.

Simchi-Levi, D., P. Kaminsky and E. Simchi-Levi (2000) *Designing and Managing the Supply Chain: Concept, Strategies, and Case Studies*, Irwin, IL: McGraw-Hill.

Slack, B. (1995) 'Containerization, inter-port competition, and port selection,' *Maritime Policy & Management*, 12(4), 293–303.

Slack, B. (1998) 'Intermodal transportation,' in B. Hoyle and R. Knowles (eds) *Modern Transport Geography (2nd edn)*, New York: John Wiley & Sons, pp. 263–290.

Slack, B. (1999) 'Satellite terminals: a local solution to hub congestion?', *Journal of Transport Geography*, 7(4), 241–246.

Slack, B. (2004) 'The global imperatives of container shipping,' in D. Pinder and B. Slack (eds) *Shipping and Ports in the Twenty-First Century*, London: Routledge, pp. 25–39.

Slack, B., C. Comtois and R. McCalla (2002) 'Strategic alliances in the container shipping industry: a global perspective,' *Maritime Policy & Management*, 29(1), 65–76.

Slack, B., C. Comtois and G. Sletmo (1996) 'Shipping lines as agents of change in the port industry,' *Maritime Policy & Management*, 23(3), 289–300.

Sminia, H. and A. Van Nistelrooij (2006) 'Strategic management and organisation development: planned change in a public sector organisation,' *Journal of Change Management*, 6(1), 99–113.

Solow, R.M. (1956) 'A contribution to the theory of economic growth,' *The Quarterly Journal of Economics*, 70(1), 65–94.

Solow, R.M. (1957) 'Technical change and the aggregate production function,' *Review of Economics and Statistics*, 39(3), 312–320.

Steinmo, S., K. Thelen and F. Longstreth (1992) *Structuring Politics: Historical Institutionalism and Comparative Analysis*, Cambridge: Cambridge University Press.

Stiglitz, J.E (2006) *Making Globalisation Work*, New York: Norton.

Stone, B.A. (1998) 'Is intermodalism sustainable?,' *Transportation Quarterly*, 52(4), 77–87.

Stubbs, P.C., W.J. Tyson and M.Q. Dalvi (1984) *Transport Economics (Revised Edition)*, London: George Allen & Unwin.

Taaffe, E.J., H.L. Gauthier and M.E. O'Kelly (1996) *Geography of Transportation (Second Edn)*, Upper Saddle River, NJ: Prentice Hall.

Taaffe, E.J., R.L. Morrill and P.R. Gould (1963) 'Transport expansion in underdeveloped countries: a comparative analysis,' *Geographical Review*, 53, 503–529.

Talley, W.K. and M.W. Ng (2013) 'Maritime transport chain choice by carriers, ports and shippers,' *International Journal of Production Economics*, 142(2), 311–316.

Taylor, T.A. (2002) 'Supply chain coordination under channel rebates with sales effort effects,' *Management Science*, 48(8), 992–1007.

Tecnologística (2008) *Indefinição prejudica setor de portos secos*, Revista Tecnologística, May 2008.

Theys, C., T.E. Notteboom, A.A. Pallis and P.W. de Langen (2010) 'The economics behind the awarding of terminals in seaports: towards a research agenda,' *Research in Transportation Economics*, 27(1), 37–50.

Toonen, T.A.J. (1998) 'Provinces versus urban centres: current developments background and evaluation of regionalisation in the Netherlands,' in P. Le Gales and C. Lequesne (eds) *Regions in Europe: The Paradox of Power*, London: Routledge, pp. 130–149.

Trade and Transport (2004) *A nova vocação dos portos secos*, Revista Trade and Transport, No. 86, July, 42–46.

Tsay, A. (1999) 'Quantity-flexibility contract and supplier-customer incentives,' *Management Science*, 45(10), 1339–1358.

UNCTAD (1994) *Review of Maritime Transport*, (http://unctad.org/en/Pages/Publications/Review-of-Maritime-Transport-(Series).aspx)

UNCTAD (1995) *Comparative Analysis of Deregulation, Commercialization and Privatization of Ports*, New York: UNCTAD.

UNESCAP (2005) *Monograph Series on Managing Globalization – Regional Shipping and Port Development Strategies – Container Traffic Forecast.* UNESCAP, Bangkok (ST/ESCAP/2398).

UNESCAP (2006) *Promoting Dry Ports as a Means of Sharing the Benefits of Globalization with Inland Locations*, UNESCAP, Bangkok (E/ESCAP/CMG(3/1)1).

Unafisco (2008) Informação dos auditors fiscais da Receita Federal do Brasil em Uruguaiana/RS, accessible at: www2.unafisco.org.br/noticias/boletins/2008/agosto/anexo_2655_uruguaiana.pdf

van Ham, J.C. (1998) 'Changing public port management in the Hamburg–Le Havre range,' in G. Sciutto and C.A. Brebbia (eds) *Maritime Engineering and Ports*, Southampton: WIT, pp. 13–21.

van Klink, H.A. and G.C. van den Berg (1998) 'Gateways and intermodalism,' *Journal of Transport Geography*, 6(1), 1–9.

Vanfraechem, S. (2012) 'Why they are tall and we are small! Competition between Antwerp and Rotterdam in the twentieth century,' in G. Harlaftis, S. Tenold and J.M. Valdaliso (eds) *World's Key Industry: History and Economics of International Shipping*, Basingstoke: Palgrave Macmillan, pp. 142–157.

Vicente, J. and R. Suire (2007) 'Informational cascades versus network externalities in locational choice: evidence of "ICT" clusters' formation and stability,' *Regional Studies*, 41(2), 173–184.

Vries, I.M.J. (2005) 'Quality of port and city requires space,' Presentation at the academic session of the *805th Anniversary of the Freeport of Riga*, June.

Wang, J.J. and A.K.Y. Ng (2011) 'The geographical connectedness of Chinese seaports with foreland markets: a new trend?,' *Tijdschrift voor Economische en Sociale Geografie*, 102(2), 188–204.

Wang, J.J. and B. Slack (2000) 'The evolution of a regional container port system: the Pearl River Delta,' *Journal of Transport Geography*, 8(4), 263–276.

Wang, J.J., A.K.Y. Ng, and D. Olivier (2004) 'Port governance in China: a review of policies in an era of internationalizing port management practices,' *Transport Policy*, 11(3), 237–250.

Wang, K., A.K.Y. Ng, J.S.L. Lam and X. Fu (2012) 'Cooperation or competition? Factors and conditions affecting regional port governance in South China,' *Maritime Economics & Logistics*, 14(3), 386–408.

Webber, M.J. (1972) *Impact of Uncertainty on Location*, Cambridge, MA.: MIT Press.

Williams, D.M. and J. Armstrong (2012) 'An appraisal of the progress of the steamship in the nineteenth century,' in G. Harlaftis, S. Tenold and J.M. Valdaliso (eds) *World's Key Industry: History and Economics of International Shipping*, Basingstoke: Palgrave Macmillan, pp. 43–63.

Williamson, O.E. (1974) 'Peak-load pricing: some further remarks,' *Bell Journal of Economics*, 5(1), 223–228.

Williamson, O.E. (1985) *The Economic Institutions of Capitalism*, New York: The Free Press.

Williamson, O.E. (2000) 'The new institutional economics: taking stock, looking ahead,' *Journal of Economic Literature*, 38(Sept), 595–613.

Williamson, O.E. (2002) 'The theory of the firm as governance structure: from choice to contract,' *Journal of Economic Perspectives*, 16(3), 171–195.

Williamson, O.E. (2008) 'Outsourcing: transaction cost economics and supply chain management,' *International Journal of Supply Chain Management*, 44(2), 1559–1576.

Wilmsmeier, G., J. Monios and B. Lambert, Bruce (2011) 'The directional development of intermodal freight corridors in relation to inland terminals,' *Journal of Transport Geography*, 19(6), 1379–1386.

Womack, J.P., D.T. Jones and D. Roos (1990) *The Machine that Changed the World: The Story of Lean Production*, New York: Harper Perennial.

World Bank (2002) *India's Transport Sector: The Challenges Ahead (Vol. 1)*, Washington D.C.: The World Bank Group.

World Bank (2007) *The World Bank Port Reform Took Kit (2nd edn)*, Washington, DC: Transport Division, The World Bank Group.

World Bank (2007) *Connecting to Compete – Trade Logistics in the Global Economy: The Logistics Performance Index and Its Indicators 2007*, Washington D.C.: The World Bank Group.

World Bank (2010) *Doing Business: Measuring Business Regulations*, The World Bank Group, Washington, DC, accessible at: www.doingbusiness.org.

World Bank (2010a) *Connecting to Compete – Trade Logistics in the Global Economy: The Logistics Performance Index and Its Indicators 2010*, Washington D.C.: The World Bank Group.

World Bank (2010b) 'India Transport Sector,' accessible at: http://web.world-bank.org/wbsite/external/countries/southasiaext.

World Bank (2012) *Connecting to Compete – Trade Logistics in the Global Economy: The Logistics Performance Index and Its Indicators 2012*, Washington D.C.: The World Bank Group.

World Economic Forum (2008) *Report on the Indian Economic Summit: Building Centres of Excellence*, Geneva: World Economic Forum.

Wu, Y.C.J. and M. Goh (2010) 'Container port efficiency in emerging and more advanced markets,' *Transportation Research Part E: Logistics and Transportation Review*, 46(6), 1030–1042.

Xiao, Y., A.K.Y. Ng, G. Yang and X. Fu (2012) 'An analysis of the dynamics of ownership, capacity investments and pricing structure of ports,' *Transport Reviews*, 32(5), 629–652.

Yan, J., X. Sun and J.J Liu (2009) 'An empirical model to assess container operators' efficiencies and efficiency changes with heterogeneous and time-variant frontiers,' *Transportation Research Part B: Methodological*, 43(1), 172–185.

Yang, S.C. (1999) *The North and South Korean Political Systems – A Comparative Analysis*, Oxford: Westview.

Yeo, G.T. and S.H. Cho (2007) 'Busan and Gwangyang: one country, two port system,' in K. Cullinane and D.W. Song (eds), *Asian Container Ports*, New York: Palgrave Macmillan, pp. 225–238.

Yeo, G.T., A.K.Y. Ng, P.T.W. Lee and Z.L. Yang (forthcoming) 'Modeling port choice in an uncertain environment,' *Maritime Policy & Management* (in press, doi: 10.1080/03088839.2013.839515).

Index

Printed and bound by CPI Group (UK) Ltd, Croydon, CR0 4YY